GENERAL FOREWORD

THERE are many men and women who, from lack of opportunity or some other reason, have grown up in ignorance of the elementary laws of science. They feel themselves continually handicapped by this ignorance. Their critical faculty is eager to submit, alike old established beliefs and revolutionary doctrines, to the test of science. But they lack the necessary knowledge.

Equally serious is the fact that another generation is at this moment growing up to a similar ignorance. The child, between the ages of six and twelve, lives in a wonderland of discovery ; he is for ever asking questions, seeking explanations of natural phenomena. It is because many parents have resorted to sentimental evasion in their replies to these questionings, and because children are often allowed either to blunder on natural truths for themselves or to remain unenlightened, that there exists the body of men and women already described. On all sides intelligent people are demanding something more concrete than theory ; on all sides they are turning to science for proof and guidance.

To meet this double need—the need of the man who would teach himself the elements of science, and the need of the child who shows himself every day eager to have them taught him—is the aim of the " Thresholds of Science " series.

This series consists of short, simply written monographs by competent authorities, dealing with every branch of science—mathematics, zoology, chemistry and the like. They are well illustrated, and issued at the cheapest possible price. When they were first published in France they met with immediate success, showing that science

†

is not necessarily dull or fenced off by a barrier of
technical jargon. Of course, specialisation in this as in
other subjects is not for everyone, but the publication of
this series of books enables any man or woman to learn,
any child to be taught, to pass with understanding and
safety the " Thresholds of Science."

MATHEMATICS

THRESHOLDS OF SCIENCE

VOLUMES ALREADY PUBLISHED

ZOOLOGY   by E. Brucker.

BOTANY   by E. Brucker.

CHEMISTRY   by Georges Darzens.

MECHANICS   by C. E. Guillaume.

MATHEMATICS

BY

C. A. LAISANT

ILLUSTRATED

DOUBLEDAY, PAGE & CO.,

GARDEN CITY   NEW YORK

1914.

CONTENTS

CONTENTS

MATHEMATICS

1. Strokes.

ONE of the first faculties which we should develop in the child, from the age when his cerebral activity wakens, is that of drawing. Nearly always, he has the instinctive taste for it, and we must encourage him in it, long before undertaking to teach him writing or reading.

With this object, we should put a slate or a sheet of squared paper in his hands as a beginning, and place a pencil (when he is cleverer, a pen) between his little fingers and make him trace strokes at first; not the classical sloping strokes, preparatory to sloping writing, but little lines following the direction of the lines on the squared paper, and very regularly spaced.

Drawing these lines first from top to bottom, then after some time from left to right, the pupil will thus make *vertical strokes* (Fig. 1) and *horizontal strokes* (Fig. 2).

FIG. 1.—Vertical strokes.

| | | | |

| | | |

FIG. 2.—Horizontal strokes.

Gradually we will teach him to draw long or short strokes, to put them between the lines of the squared paper, to draw new ones from them which are oblique, in every possible direction. Then we will make him form figures composed of groups of long or short strokes. We will say something about that below.

Later, we will make him draw figures or begin curves, either with instruments (ruler, set square, compass) or freehand. These exercises, which develop skill of hand and straightness of eye, should never be left off while the educative period lasts. We only speak of them here in so far as they are indispensable for what will come after: but, even from this point of view, we must insist on the fact that they should be suggested and never forced. If they cease to be a game, the object will be lost. Let the child scribble on his slate and spoil some sheets of paper; help him with your advice, which he will never fail to ask; but when he has had enough, let him do something else. That is a condition which is absolutely necessary to develop the spirit of initiative in him, to keep up his natural curiosity and to avoid fatigue and boredom.

It would need a whole book to deal with this first teaching of drawing, on which I have been obliged to say a few words; others would be needed on writing and on reading, which should only come afterwards and are outside my subject. But all these teachings, applied to childhood, should always be inspired by the same fundamental principle, that is, to keep the appearance of games, to respect the child's liberty and to give him the illusion (if it is one) that he himself discovers the truths put before his eyes. As to the age at which this first mathematical initiation should commence, starting with that of drawing, and then running parallel with it, there is no absolute rule to be laid down. But we can say that as a rule it is very rare if a child of three and a half to four years old does not already show a taste for handling a pencil: and I assert that at ten or eleven, it should be easy to have taught him all the matters explained in what follows, if his brain is normally organised.

More than one may find pleasure, after some years, in taking up this little book which is not meant for him now. His mind, perfected by further studies and ready at conscious reasoning, will certainly find in it matter for useful reflection.

To finish with these generalities and not to repeat myself unnecessarily, I must point out to families and teachers who are to be my readers the greatest snare to be avoided in the first teaching of childhood : this is the abuse of exercise of the memory, still so general in actual practice, and so pernicious. By teaching words to a child and making him repeat them, we deform his brain, we kill his natural gifts, and bring up generations of beings without initiative, without curiosity and without will, enfeebled, depressed, and stuffed up with formulæ which are not understood.

If you love your children, if you love those confided to you, if you wish them to become good and strong, go back to the principles of the great minds and hearts who were named La Chalotais,[1] Frœbel [2] and Pestalozzi.[3] If the earth was peopled with reasonable beings, these benefactors of humanity would have their statues in every country of the world, and their names would be graven in letters of gold in every school.

[1] La Chalotais, French magistrate, born at Rennes (1701—1785), author of " Essay on National Education."
[2] Frœbel, German pedagogue, born at Oberweissbach (1782—1852), founder of the " Kindergartens."
[3] Pestalozzi, Swiss educator, born at Zurich (1746—1827) ; his method served as a basis for Fichte, as a means of raising Germany.

2. One to Ten.

Once the habit has begun of drawing strokes regularly—
and quickly enough—he will learn to count them as he
makes them, pronouncing the names, one, two, three,
four, five, six, seven, eight, nine, ten, successively.

Then he will make groups of strokes, separating them
from one another by spaces, and he will have (Figs. 3 and
4) diagrams which he will read :—

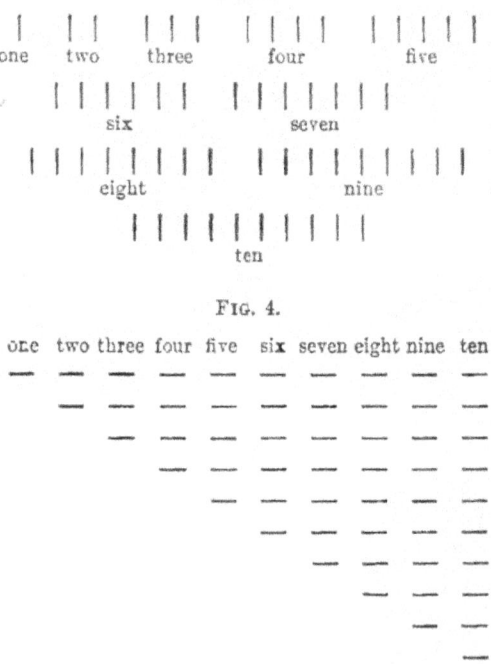

FIG. 3.

FIG. 4.

One, two, . . . ten vertical strokes, for Fig. 3 ;
One, two, . . . ten horizontal strokes, for Fig. 4.

Then he will put down groups of beans, of grains of corn, of counters, and of any other things, and they are to be counted :—

One, two, . . . ten beans, grains of corn, etc.

We will now suppose that these objects are replaced by sheep, dogs, men, etc., and after these exercises are repeated often enough and are familiar to the child, we can tell him that the expressions he uses, for instance, three strokes, six grains of corn, eight sheep, are called *concrete numbers*.

Having considered a group of five strokes, another of five beans, another of five counters, having imagined another of five dogs or five trees, we will tell him that in these different cases he is always saying the same word *five ;* we will tell that this word, without anything else, represents what is called an *abstract number*, and that he can use it to denote any other group of five things ; donkeys, horses, houses, etc.

It will not be long before the child can count without any hesitation from one to ten in any things at all. It will be well also to accustom him as soon as possible to grasp at a glance the number of things shown to him quickly, without having to count them one by one : to do this, he must start with very small numbers and go on gradually.

3. Matches or Sticks; Bundles and Faggots.

Beyond the different things just mentioned, to help the child to understand the idea of concrete numbers, which can be infinitely varied, there are others which we can hardly recommend too highly, whose use is, in our opinion, indispensable. These are little wooden sticks, exactly like ordinary wood matches except that they have no inflammable chemical preparation. We will sometimes call them matches, because of this resemblance, and these matches—which do not strike—can be considered models of the strokes drawn on the slate or in the copy-book. They should all be the same length.

Having a heap of these sticks before him, and knowing quite well how to count up to ten, the child will put aside ten, one after the other, and will put them together in a very regular little bundle, and surround them with one of those little rubber bands which are so convenient and so widely used.

We will show him then that this *bundle* containing ten sticks can be called a " ten " of sticks.

Then he will make up a fair number of similar bundles. We will see that he has not made a mistake : if he has we will make him put it right.

Then showing him two bundles, we will tell him that in these two bundles taken together, the number of sticks which we will show him by untying the bundles and tying them up again is called *twenty*, and that thus :—

> *One* bundle is *ten* sticks,
> *Two* bundles are *twenty* sticks.

Then taking three, four, . . . nine bundles, and doing the same thing, we will show him that

Three	bundles are	thirty	sticks
Four	,,	,, forty	,,
Five	,,	,, fifty	,,
Six	,,	,, sixty	,,
Seven	,,	,, seventy	,,
Eight	,,	,, eighty	,,
Nine	,,	,, ninety	,,

Having learnt all this, to conclude we will take ten bundles, and we will put them together by a larger rubber band, which will give us a *faggot*. We will then explain that a faggot is *a hundred* sticks, that the number of sticks in a faggot is called a *hundred :* he will verify that as ten bundles make a faggot, *ten tens* are *a hundred*.

4. One to a Hundred.

Taking a handful of sticks at random (less than a hundred) we will tell the child that we are going to count them with him. To do this, he must make bundles, as many as possible; he will come to a point at which he will not have enough sticks left to make a bundle. Then, putting all the made-up bundles at his left, and the sticks left over at his right, we will make him say the two numbers separately; then, putting them together, he will have named the number of sticks we gave him.

For example, if he has made *three* bundles, and *eight* sticks are left over, he will say, looking to the left, " thirty "; looking to the right, " eight "; then, without a stop, " thirty-eight."

Having repeated this exercise very often, with groups of sticks taken at chance, we will take a faggot to pieces, and propose to count the sticks successively one by one. We will begin by counting one, two, three, . . . to ten. Then, having a bundle, we will place it on the left (without even needing to tie it) and go on, saying :—

One-ten ; two-ten[1] *; thirteen; fourteen ; fifteen;*
Sixteen ; seventeen ; eighteen ; nineteen.

At last a fresh stick finishes a second bundle, which we put on the left, by the side of the first, saying *twenty ;* we go on in the same way to the ninth bundle, then to the ninth stick left, which we touch saying *ninety-nine ;* finally we remove the last, finishing the tenth bundle, which we put on the left, by the side of the nine, saying the word *hundred.*

There is nothing to prevent us telling the young pupil then that we have just taught him *numeration* from one

---

[1] Here we must not say "eleven, twelve." These names can be learnt without any trouble at the right moment. It is useless to load the memory with them now. Even if the child in his logic says " three-ten, four-ten," etc., there is no need to correct him yet.

to a hundred ; we can even tell him that when he says
" seventy-three " matches or sticks, he is performing *spoken
numeration,* and that, when he puts seven bundles to the
left and three sticks to the right, he is performing
numeration by figures. He will be all the more flattered
to feel himself so wise since he does not yet know how to
write a letter or a figure, or to read, b, a, ba. But he
draws strokes, he has eyes, he uses them to see, and
begins to understand what he sees and what he is doing.

So now we know how to count sticks from one to a
hundred. We must accustom ourselves to count any
other things in the same way, and then to count them in
our heads at once without having them before our eyes.
That is the beginning of *mental calculation,* so important
in practice and so easy to make children practise from
the earliest age, if we begin with very simple things and
go on gradually.

This is not yet all ; starting from one, we must become
accustomed to count by twos :—

One, three, . . . to ninety-nine,

and explain that all these numbers are *odd numbers.*
We must do the same, starting from two :—

Two, four, six, . . . to a hundred,

and we will have the *even numbers.* We will then become
accustomed to count by threes, by fours, starting at first
from one, and then from any number.

All these exercises are to be done first with things—
preferably sticks—then mentally.

In short, this manipulation of numbers, from one to a
hundred, can be varied indefinitely, for we must not fear
prolonging it as long as it does not become tedious and
as long as it interests the child. It will be well to come
back to it from time to time, even when he has penetrated
a little further in his introduction to science.

5. The Addition Table.

Let us arrange on a table, from left to right, one, two, . . . to nine sticks, separating these nine groups from one another. Below the single stick, let us put two and make a column, starting two, three, to ten. A second column made in the same way will contain three, four, . . . one-ten sticks; and going on in the same way we will have nine columns; the last group of the ninth column will be of eighteen sticks.

Now is the moment to come back and use our skill in drawing and our great aptitude in tracing strokes. Only, as it is troublesome to draw the ten strokes representing the sticks in a bundle, we will make a picture of a bundle by a thick stroke, stronger than the others, made of two parts, H, with a little bar to recall the presence of the rubber band. Thus we have now begun to know how to write numbers with strokes, and in this way copying the figure as we are going to

FIG. 5.

explain, we will have Fig. 5, at least partly. To finish it, we will put one, two, . . . nine strokes on the left of the two, three, . . . ten of the first column; finally, we will separate this new column by a vertical line from the rest of the figure, and we will also separate the first row by a horizontal line.

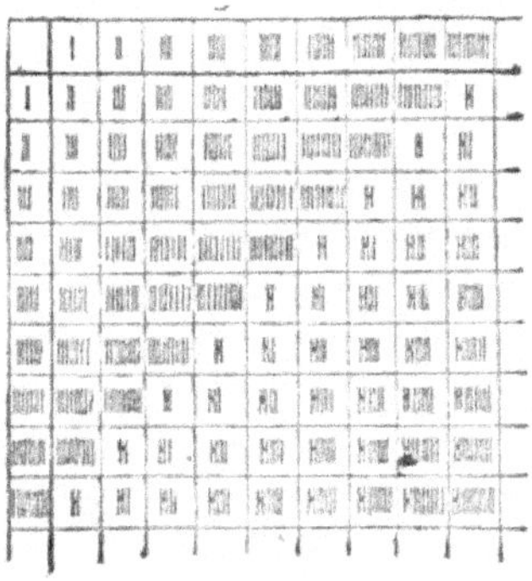

The figure we have thus obtained is an *addition table ;* we will soon see why it is so called.

It lends itself to several interesting remarks which the maker will partly find out. First, all the numbers in the same slanting line, rising from left to right, are the same; further, all the numbers, read from left to right in a horizontal line, or from top to bottom in a column, are the numbers counted one by one; finally, the numbers in the same slanting line going down from left to right, if read downwards, are the numbers counted two by two. These are sometimes even, sometimes odd.

Nothing stops us from reading all these numbers in the reverse order, which will teach us to say the numbers fluently, one by one, or two by two, in the direction opposite to the natural order. There is still a very important exercise of which we have not yet spoken, and which we will now begin with small numbers; this will not present any serious difficulty.

6. Sums.

Let us take two heaps of beans (or other things) and count them both. If we put them all into one heap, how many shall we have? For this, we have only to count again the heap made by mixing the other two. But this would be lengthy and wearisome, and it would be time lost.

We will explain that there is a quicker way of getting the result, that we will do it by an operation called *addition*, and that the number of things in the big heap, which we want to find, is called the *sum* or *total*.

Taking numbers smaller than ten, and looking again at Fig. 5, we notice that it gives all the sums of two heaps, and we will invite the child to try and find this out. We will do this, repeating these exercises as often as possible and making him count the sum itself when he does not find it.

Even before this addition table is completely fixed in the memory, we will take any two numbers—chosen so that their sum is less than a hundred—and we will count them both separately. We will then represent them by sticks; suppose they are thirty-four and twenty-three.

The first number is made up of three bundles and four sticks; the other, of two bundles and three sticks, is placed below—bundles under bundles, on the left, sticks under sticks, on the right.

We will then ask the child to say how many are made by four and three sticks; he will answer "seven," helping himself if necessary by the addition table, and he will place seven sticks a little lower down. So, how many do three and two bundles make? Five bundles, which we will place below the bundles. We have thus the total, five bundles, seven sticks, or fifty-seven sticks.

We will begin again with other numbers, taking those where there are only bundles and no sticks, like sixty, twenty, eighty; others where there are no bundles, numbers less than ten; but so that each sum of bundles or sticks is always less than ten.

When we have reached this point, we will take other numbers where this is not so; for instance, forty-nine and twenty-five.

The operation is performed thus :—

Four bundles	Nine sticks
Two bundles	Five sticks

We have then nine and five, or fourteen sticks; this gives us a bundle—which we put under the bundles—and four sticks. Then counting the bundles, beginning with that we have just made, we have one and four, five; five and two bundles, seven. The total is thus seven bundles and four sticks, or seventy-four.

This exercise should be repeated, renewed with different examples, so that it will interest the child without boring him.

Then, coming to additions of several numbers, we will proceed in the same way (always arranging that the total is less than a hundred), and we will see that thus we find the number formed by joining several heaps, when we know the number in each heap.[1]

Repeat these exercises on a crowd of examples as long as they do not cause fatigue or boredom. If the child seems to be at all troublesome, the punishment should consist in a threat—carried out for several days—not to go on showing him the game of sticks, counters, etc., which he has begun to learn. Let this device be used with some skill and we will see that it is not hard to lead the offenders back to their studies by their own wishes. Only, do not pronounce in their ears the unfortunate word " study," which might frighten them.

7. Subtraction.

I have a big heap of counters, eighty-seven let us say ; I pick up or take away a few which I count : I find there are twenty-five. How many are left ? To find that is to do a subtraction ; the result is the remainder or difference ; we notice that if we put back the remainder to the number from which we have taken it we make once more the big heap, by which I mean the number from which I subtract. To find the difference, first let us write the larger number, eighty-seven, with little sticks :—

Eight bundles　　　Seven little sticks,

and, underneath, the smaller one, twenty-five :—

Two bundles　　　Five little sticks,

taking great care to put the bundles to the left, the little sticks to the right, and to put the little sticks under each other and the bundles under each other. From the larger number I take away five little sticks ; I shall have

---

[1] These exercises will oblige the child to learn, beyond the addition table, how to add quickly a number less than ten to a number less than a hundred ; for instance, sixty-eight and five, seventy-three. With a little patience this result will be obtained by practice quickly enough.

two left. I take away two bundles, and six will remain. Then the remainder will be

Six bundles   Two little sticks,

or sixty-two little sticks.

Now we are able to subtract (only doing it with numbers less than ten), since we have taken away five from seven, and two from eight. But it isn't always so easy! For instance, the big heap may count up to fifty-two, and what we wish to take away may be eighteen, which is certainly less. Arranging them as before we have

Five bundles   Two little sticks;
One bundle   Eight little sticks.

We cannot take away eight sticks from two. So from among the five bundles we take one which we put to the right with the two sticks. Whether we undo it or not we can see very well that now we have ten-two sticks to the right, and only four bundles to the left instead of five. Now from the ten-two little sticks at the right we shall take away eight. We shall have four left; from the four bundles at the left we take away one: there are three left. The remainder then is

Three bundles   Four little sticks,

or thirty-four.

We must know how to subtract a number less than ten from a number larger than ten, but which will be always less than twenty. By multiplying these exercises many times, and by varying them as much as possible, this subtraction (which has to be learnt) will be remembered by the pupil; but above all, do not cause them to be learnt by heart and recited. Do them all the time instead: this is much more effective. We must take care always to take for the larger number one less than a hundred, because at present we cannot count beyond that.

8. Thousands and Millions.

Up to now we know how to count up to one hundred. It is a big number if we consider the age of a person in years ; a man aged one hundred is very old, and centenarians are very rare. But it is a very little number if we count grains of corn ; a heap of a hundred grains of corn is not at all big, and would not be sufficient to feed a child for a day, so that it is impossible to stop there, and we shall be obliged to climb still further up the ladder, which will not, however, be difficult.

We have got up to one hundred by grouping the little sticks in bundles of ten, and grouping ten bundles of those in a faggot, which contains one hundred little sticks. Let us put together ten faggots in a box, then with ten such boxes let us form a bale, ten bales can be put together in a basket, with ten baskets we will make a case, with ten such cases we will make a wagon-load, with ten wagon-loads a car-load, and with ten car-loads a train.

Going all over this, we are going to give the names of the numbers that we obtain in this manner.

A match or a stick is what we will call a *simple unit*.

In a *bundle* we have *ten* matches, or one set of ten.

In a *faggot* of ten bundles, *one hundred* matches, or a set of *one hundred*.

In a *box* of *ten* faggots, a *thousand* matches.

In a *bale* of ten boxes, *ten thousand* matches or one *ten* of *thousands*.

In a *basket* of ten bales, *one hundred thousand* or one *hundred* of *thousands*.

In a *case* of ten baskets, a *million*.

In a *wagon-load* of ten cases, *ten millions*, or one *ten* of *millions*.

In a *car-load* of ten wagon-loads, *one hundred millions*, or a *hundred of millions*.

In a *train* of ten car-loads. *one thousand millions*.

We might go on as long as we liked, but the number—

a thousand millions—at which we have arrived, is large enough for ordinary use. We should have an idea of its size if we placed ordinary wooden matches, one after another, to the number of one thousand millions; the total length would be considerably more than the circumference of the earth.

Trying to count a thousand million matches one by one, supposing that we took a second for each, and occupying in this counting ten hours a day, we would take more than seventy-six years. This would perhaps be rather long, not very amusing, and only slightly instructive.

No, if we want to count a big heap of little sticks, we will make bundles of them, and we will put to the right the little sticks that are left over; once the bundles are made, perhaps there may be *three* little sticks. We will now make faggots with our bundles, making them up in tens; suppose there remain *eight* bundles, we will place them to the left of the three little sticks; we will count our faggots in tens to make boxes of them. *Five* faggots are left, we place them to the left of the eight bundles and counting our boxes we find *six* of them. We put them to the left of the five faggots and we have thus the number of little sticks: *six* boxes, *five* faggots, *eight* bundles, *three* little sticks; or six thousand five hundred and eighty-three little sticks. With nothing but bundles and faggots we shall be able to count up to a thousand, and form all the numbers up to that, never forgetting that faggot, bundle, single little stick mean respectively

a hundred, ten, one, little sticks.

If in the number that we wish to unite there are no single little sticks, or no bundles, that will not make any difference. For instance,

Eight faggots, six bundles,

Will contain eight hundred and sixty little sticks, and

Five faggots, three little sticks

Will contain five hundred and three.

It will be necessary to have numbers under a thousand formed like this, and to have performed many additions and subtractions, exactly as it has been shown before, but extending the method of procedure as far as faggots instead of keeping to bundles.

It is advisable to notice that we often meet the same numbers, ten and hundred, or tens and hundreds. Thus:

Little stick			one
Bundle	} mean	{	one ten
Faggot			one hundred
Box			one thousand
Bale	} ,,	{	one ten of thousands
Basket			one hundred of thousands
Case			one million
Wagon-load	} ,,	{	one ten of millions
Car-load			one hundred of millions

A number of thousands or of millions will be reckoned then as if we were counting simple little sticks from one to a thousand. Thus :—

Three car-loads	Two wagon-loads	Seven cases
One basket	(no bales)	Nine boxes
Four faggots	Five bundles	

will be a number of little sticks which will express

Three hundred and twenty-seven millions
One hundred and nine thousand } little
Four hundred and fifty } sticks.

We could have some of them counted like that, but without insisting upon the large numbers for the moment, and applying ourselves particularly to the bundles, and faggots, going no further than the boxes at the very most.

Always, in what has gone before, we have taken care to place the little sticks—single ones—to the right, the bundles—tens—to the left; the faggots—hundreds—to the left of the bundles, and so on. Strictly speaking, we ought to notice that this is unnecessary, but it is most convenient, and it is a good thing to keep to this arrangement always, because the calculation is done in order.

A little later, the child, having been accustomed to this habit, will find it natural ; this will be valuable, as it will be indispensable when calculating. To form effectively, with little sticks, all the numbers of which we have just spoken, and of which it is a good thing to speak to settle the child's mind, we must find something rather heavy, and very easy to place on a table, or on a sheet of paper, even before making it up into car-loads. We are going to see now how we can simplify things, and show the young mathematician—who cannot either read or write fluently—that it is perfectly easy to manage with his fingers the enormous numbers with which he has to deal.

9. Coloured Counters.

It is very disagreeable to be so encumbered, as soon as we want to count a thousand matches, by our bundles and faggots. As we know already that numbers can be represented by any means, let us replace our matches by white counters. That does not alter our calculations nor the manner of doing them. Now, let us change our bundles for red counters ; they will be really more convenient to manage, and we can always replace a red by ten white ones if it is necessary. To carry it still further, instead of faggots we will put orange counters ; instead of boxes, yellow counters ; instead of bales, green counters; instead of baskets, blue counters ; instead of cases, indigo counters ; instead of wagon-loads, violet counters; instead of car-loads, black counters ; and finally, instead of trains, long counters—white ones.

The objects and the numbers correspond in this manner :

Matches { Trains, car-loads, wagon-loads, cases, baskets, bales, boxes, faggots, bundles and matches.

M.   C

Counters ｛ Long, black, violet, indigo, blue, green, yellow, orange, red, white.

Numbers ｛ Thousands of millions, hundreds of millions, tens of millions, millions, hundreds of thousands, tens of thousands, thousands, hundreds, tens, units.

Nothing hinders us then from writing all the numbers that we require up to a thousand millions, and even further than that, with our little counters, without being compelled to use cases, car-loads, and even trains; and it is equally in our power, if that will interest us, to add and subtract. It will be necessary, though, always to remember that a red counter is equal to ten white, an orange counter to ten red, and so on to the end.

It might seem that for white counters, we might put cents, then replace the red counters by dimes, and continue like that; but that would become awkward and cumbersome, and we would be obliged to have a nice little fortune; because then, to represent thousands of millions, we would have to use coins each worth ten million dollars. Money of that value is not coined, it would be difficult to handle, and it will be decidedly better to continue to use long white counters to represent thousands of millions. It will also be more economical! Always, as we go higher, we will put our counters carefully in order, beginning at the right.

Long.	Black.	Violet.	Indigo.	Blue.	Green.	Yellow.	Orange.	Red.	White.

Simply by looking at each place we know what colour ought to go there according to its place, starting from the right.

10. Figures.

We know now how to write all the numbers, at least as far as thousands of millions—and it would be easy to go on much further—with our round counters of various colours, and the long white counters. To do that, we must put in each of the places where we see the white or red counters, etc., or units, and tens, etc., a number of counters which is always less than ten.

If by any means we could avoid reckoning these counters every time, it would be much more convenient. By this time the pupil has commenced to write a little, and we can exercise him in tracing the characters which will represent the first nine numbers, which we shall need—characters which are called "figures." These are :—

One two three four five six seven eight nine
1   2   3   4   5   6   7   8   9

Whether with a pencil or with a pen, he will become accustomed to form them correctly without any flourishing, with a single stroke, except perhaps the figures four and five, which require two, making use of ruled slates or ruled paper at first, so that the figures may be all the same height. This is of the highest importance for the future practice of calculations. Here is the type to which we must keep :—

1 2 3 4 5 6 7 8 9

Just for curiosity we may notice here that all the figures, according to some old authors, owe their origin to the figure   ⊠   (see below); but this is not well authenticated.

The important point is to make the pupil change the numbers from little sticks into counters, and from counters into figures, not taking very large numbers, especially to begin with. We will notice that there is no necessity to make our figures of different colours, because the place which they occupy tells us quite easily whether they represent simple units, tens, hundreds, etc., or counters coloured white, red, orange, etc., or, again, little sticks, bundles, faggots, etc.

But now comes an important observation. If there are no counters at all of a certain colour, we don't put anything there. There is only the place in the row to distinguish the figures; if we put nothing there it would mix everything up, because we ought to leave a space always the same size, that of a figure, and we are not clever enough to write always so regularly. Besides, if the space happened to be in the unit place, how could we know the meaning of the last figure to the right? To escape all these difficulties, we put, in empty spaces, a round character, 0, that we call, " zero "[1] which has no value but fills up the place. Zero is a good modest servant who guards the house, and who says to you: " There is nobody here. As for me, I do not count, I am nothing, but I hinder anybody from coming in."

From now onward we can teach the pupil to write quantities of numbers often using zero or " nought," and varying exercises of this kind. If there are several pupils, one might exercise them together, rousing

---

[1] The inventor of the zero is not known, but this clever idea seems to be of Hindoo origin.

their emulation a little, leading them on, more and more, to read and write quickly and correctly, and tell them at the end of the calculation that they now understand *written numeration.*

Having arrived at this stage, it is a good thing to go back again to the examples of addition and subtraction that we have been able to express with little sticks, or counters, making use now of figures. But there will be various very useful observations to make which formerly would not have been of any service. One of them, in doing addition, consists in accustoming the pupil to speak as little as possible, never to say, for instance, "I put down such and such a figure, and I carry such and such a number." It will be sufficient to make oneself understood to give the example of addition shown here :

$$
\begin{array}{r}
3087 \\
6944 \\
560 \\
208 \\
29 \\
2004 \\
\hline
12832
\end{array}
$$

Which ought to be translated then in spoken language : " The figures 7 and 4 = one-ten, and 8 = nine-ten, and 9 = twenty-eight, and 4 = thirty-two " (we write 2 without saying anything); then we add, " I carry 3, and 8 = one-ten, and 4 = five-ten, and 6 = twenty-one, and 2 = twenty-three " (we write 3); " I carry 2, and 9 = one-ten, and 5 = six-ten, and 2 = eight-ten " (we write 8); " I carry 1, and 3 = 4, and 6 = ten, and 2 = two-ten " (we write 2, then 1 to the left of that); and we read the total, "two-ten thousand, eight hundred and thirty-two."

A second remark applies to the practice of

subtraction, when there is in the greater number, in a certain row, a figure less than the one below. Let us go back to the example in section 7; from 52 we have to take 18.

52	4	12
18	1	8
34	3	4

What we have done with our little sticks is shown above. But we must not write anything else than 52 and 18 before the result of the operation, and it might very easily happen that we forget that we have taken away a ten from the top, and there only remain four tens instead of five. For the future we proceed in another way, noticing that to take 1 from 4 is the same thing as taking 2 from 5. And we would say, "8, from two-ten, equals 4. I carry 1, 1 and 1 make 2, 2 from 5 equals 3." Then instead of saying "1 from 4," I say "2 from 5," which leaves 3; so we get into the habit of carrying one each time that we have previously added a 10 to the figure from above.

Many exercises in addition and subtraction ought to be carried out like this. The child will interest himself in them, but do not try to prove anything to him. If he seems at times puzzled, take him back to his sticks or to his counters; and try only to give him practice in calculation and not to make him learn words that he does not understand. If any observations come into his mind, and he tells you of them, listen to him with great attention. Do not be afraid of going back from time to time, in order to accustom him to compare his numbers written in figures with collections of little sticks, counters, or any other objects. And, above all, do not make the lesson too long; do not let his interest flag or fatigue to overcome him; this is the teacher's deadliest scourge.

If you think it convenient, you can, from this time,

though there is no hurry about it, initiate the pupil into the ordinary names for the numbers 11 and 12.

11. Sticks End to End.

Let us make use once more of the little sticks that we have already employed ; we will imagine that we have, say, three heaps in which there are 5, 3 and 4 little sticks. If we put all the little sticks one after another in the same direction, the length of this row will be twelve little sticks, that is to say, it will give the sum of the numbers represented by the three heaps.

We should arrive at the same result if we replaced the little sticks belonging to the first heap by a straw equal in length to the five little sticks ; those from the second heap by one as long as the three little sticks ; and those from the third heap by a straw measuring the length of the four little sticks.

If, instead of these very little numbers we took larger ones, and if, instead of three numbers we took as many as we liked, all that we have just said would repeat itself. The straws would be longer; there would be more than three straws; that is all the difference.

We prove in this manner that any number whatever can be represented by a straw of suitable length, and that to find the sum of several numbers we have nothing to do but lay end to end, one after the other, the straws which represent these numbers. The length of the line of straws thus obtained will be the desired sum.

12. The Straight Line.

The straws of which we have just been speaking in the foregoing operations ought always to be placed in a *straight line* immediately after each other. But what is a straight line ? We have an idea of it by the stroke which a very fine-pointed pencil makes moving along against a ruler held horizontally, or by an extremely fine thread, a hair, for instance, held between two supports. This general idea is sufficient for us; we know quite well that if the ruler were longer, the sheet of paper wider, we would be able to draw our straight line further, either to one end or to the other : and as there is no need ever to stop, we under-stand that the straight line is, so to speak, an indefinite figure. We will never make use of it any further than the limit that we require ; but this limit can be as distant as we wish.

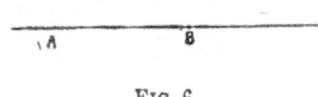

FIG. 6.

If we take a straight line (Fig. 6) and mark off a point A, and another point B, the portion of the straight line AB comprised between these two points is what we call a segment of a straight line. The straws which we made use of just now can be laid on the segments of the straight line, and the length of these straws is the same as the segments upon which they are laid.

Thus (Fig. 7), to return to the example (section 11), let us take a straight line upon which we mark a point O, no matter where : starting from this point, let us take a segment OA, which is of the same length as our first straw, five little sticks; starting from A, let us take a segment AB, having its length the same as that of

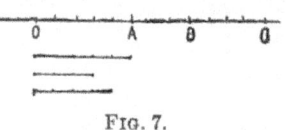

FIG. 7.

the second straw, three little sticks ; then starting from B, another, BC, of which the length equals that of the third straw, four little sticks. The segment OC will be the length of twelve little sticks ; the sum of 5, 3 and 4. Whether we say that we add numbers, straws, segments of a straight line, it is always the same thing, the addition is made by laying the straws, or the segments end to end, one after another.   This operation must necessarily be done, if with segments, always in the same direction; we will suppose it be invariably from left to right.

In Fig. 7 we can thus go on adding as far as we like to the right of O, but never to the left.

13. Subtraction with Little Sticks.

It is not any more difficult to subtract than to add, by the aid of little sticks.   Suppose, for example, that we want to take 4 from 11.   We will lay 11 sticks end to end in a straight line; then, beginning at the end to the right of this row, we take away 4 little sticks; a row of 7 little sticks remains; 7 is the difference between 11 and 4.

If we begin by putting a straw as long as 11 little sticks, it would seem as though we were obliged to cut off an end of it equal in length to the four to make the difference. But there is another way which we will understand at once, using segments instead of straws.

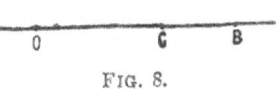

FIG. 8.

Let us lay out on a straight line, beginning from the point O, a segment OB the length of 11 little sticks.   Starting from B, let us take a segment the length of 4, but instead of supposing it traced from left to right, let us take it, on the contrary, from right to left.   The segment OC will represent by its length the difference 7.

We go over this a few times, saying that to add several segments we must lay them end to end in the same direction;   and to subtract one segment from another we must lay them end to end in the same manner, but in the opposite direction.

These things, besides being easy, are perfectly under-standable; it suffices to vary the examples a little from time to time to interest the pupil; we must not be afraid of making him manipulate little sticks (very simple to procure) as much as possible, and reproduce his exercises on a slate or a paper.

We are now going to enter the regions of higher arithmetic. If he tends to be puffed up, repress this display of pride, hinting to him, first, that Algebra is one of the easiest parts of Mathematics, and second, that he does not know anything and is not learning anything now, except games which will be useful to him later, when he remembers them.

14. We begin Algebra.

Up to now we have learned to add, giving the sums, and to subtract, giving the difference. For example, the sum of 8, 5, and 14 is 27. We have imagined a sign or symbol (+) which represents addition, and which expresses *plus*, and also a symbol (=) which expresses *equal to*. So that in this manner the result which we have just recalled might equally be written

$$8 + 5 + 14 = 27$$

and would read 8 plus 5 plus 14 equals 27.

Similarly, for subtraction, we make use of a symbol (−) which expresses *minus*, and if we write $7 - 5 = 2$, that would read, 7 minus 5 equals 2, which means that in subtracting 5 from 7, we obtain 2 as the difference.

All operations of this nature can be worked by means of straws, or segments, as we have seen before. Thus, looking at Fig. 7 we see that it denotes

$$5 + 3 + 4 = 12$$

and that it can be written just as easily

$$OA + AB + BC = OC$$

Fig. 8 denotes

$$11 - 4 = 7.$$

We can amuse ourselves by doing this in whatever manner we like, and in expressing our working under these different forms.

We understand that in the place of 8, 5, 14, or of 5, 3, 4, in the preceding examples we could put any other numbers we like; if we call them A, B, C, writing $A + B + C = S$, we will always express the sum of three numbers; this sum would be 27 in the first example, 12 in the second.

In the same manner $A - B = R$ shows that the difference obtained in subtracting B from A is equal to R. For instance, in Fig. 8, $A = 11$, $B = 4$, and $R = 7$.

It is often very convenient to show our work by signs, and to replace numbers by letters. It is well to accustom ourselves to this early, because it will be most useful in the future, and save much trouble. We must also know what it means when we put something between brackets thus

$$(\quad) + (\quad) \text{ or } (\quad) - (\quad)$$

This simply means that we should replace each bracketed term by the result which it gives. For instance,

$$(A - B) - (C - D) + (E - F)$$

if $\qquad$ A, B, C, D, E, F,

are replaced by $\qquad$ 10, 2, 9, 6, 7, 5,

would express $(10 - 2) - (9 - 6) + (7 - 5)$,

or $\qquad$ $8 - 3 + 2$, that is to say 7.

All these ways of expression are sometimes called algebraical. But the words themselves are of little importance, it is their meaning which counts.

What follows will show us something fresh. When we are adding numbers we can go on indefinitely, for instance, with several heaps of beans we can always make them into a single heap. In other words, it is always possible to add, and we can express the addition in figures, in counters, in matches, in little sticks, in straws, in segments of a straight line, just as we please.

It is not the same with subtraction. If I have a heap of seven counters, for instance, from which I wish to take away 10, the thing, as we have already remarked, is manifestly impossible.

However, if we return to what has been previously said, as shown in Fig. 8, we shall be obliged, to subtract by means of straws, or of seg-

Fig. 9.

ments of a straight line, to lay out (Fig. 9) on a straight line a segment OB equal in length to 7 matches, then from B lay in the contrary direction, that is from right to left, a segment whose length is the number to subtract; now this is always possible; and Fig. 9 demonstrates it, supposing, as we have made it, that the number to subtract is 10; we obtain thus, the length BC being 10, a point C, and we have for remainder the segment OC; only, the point C is no longer to the right of the point O; it is to the left; the segment OC is directed from right to left, and its length is equal to 3.

Such a number is said to be negative; we write it — 3, we call it " minus 3 "; and it would be correct to say $7 - 10 = -3$.

This creation of negative numbers makes all the subtractions possible which were not so with ordinary numbers, which we call by contrast positive numbers.

In Fig. 10 all the part to the right of point O represents the domain of positive numbers (first arrow); all the part to the left (second arrow), represents the domain of the negative numbers; and the total of the two arrows, comprising the straight line in its entirety, in the two directions, represents the domain of Algebra.

It will be necessary now, when we wish to express numbers by straws, or segments, to pay attention to the direction of these segments, or to the sign of the number;

thus (Fig. 9) OB will be a positive segment, representing the number 7 ; OC will be a negative segment representing the number — 3, itself also negative.

This compels us to consider, for fear of making an error, which of the two ends of a segment we will call the beginning, and which the end ; and the direction of the segment will be always that which starts

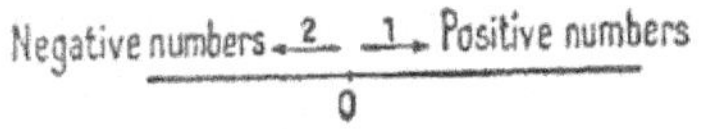

Negative numbers ←—2— —1—→ Positive numbers

0

FIG. 10.

from the beginning to go towards the end. When we write the segment AB, that will always mean that A is the beginning and B is the end. We shall be obliged to change slightly the appearance of our little sticks. It will be quite easy to blacken them slightly at one end by dipping them in Indian ink, a harmless dye ; the black part will then always represent the end. So that, placing three matches in a row, the black end towards the right, we shall express the number + 3 ; placing two of them in a row, the black end to the left, the number — 2 is shown, and so on.

It will always be correct to add one number to another by laying end to end, in the proper direction, the segments which express them. For instance, to add 11 and — 4, we will take a segment OB of the length of 11, directed from left to right, and afterwards a segment BC the length of 4 directed from right to left. Now Fig. 8 is just what we have done to obtain the difference 11 — 4. We can, therefore, write $11 + (- 4) = 11 - 4 = 7$; and subtractions thus take us back to additions.

Exercises on the negative numbers can be varied as much as we like, and will be quite easy to do with sticks blackened at one end. We can also make straws as long as several sticks, and blacken them in the same manner, to show which is the end. It is easy to get accustomed to this simple and necessary idea of the sign or the direction of number.

Besides, if the negative numbers seem puzzling at

first, a little reflection will find an altogether natural explanation. At the first blush it seems as if there could not be any number less than nothing, that is zero. However, in ordinary speech we say every day that the thermometer registers so many degrees below zero. When we wish to show the height above sea-level of any point, we understand that if this point were at the bottom of the sea, it would be below zero.

If starting from home I want to calculate the distance that I shall go in one direction, and if I walk in exactly the opposite direction, I know perfectly well that I cannot use the same number to express two opposite things.

A man without any fortune, but who owes nothing, is not rich ; but, if he has no fortune and has debts, we can say that he has less than nothing ; his fortune is negative.

A cork has a certain weight ; if we throw it in the air, it falls ; plunge it into water, and let it go, it rises ; its weight has become negative, in appearance at least.

Briefly, negative numbers, far from being mysterious in their character, adapt themselves, in the most natural fashion, to all quantities, and there are some which, from their nature, can be measured in two ways opposed to each other, such as hot and cold, high and low, credit and debit, future and past, etc. By means of concrete examples, we can make these simple ideas sink into the mind of very young children, for everything we have said is extremely simple and easy of comprehension.

The pupils will be interested if we continually accompany our explanations with examples carried out by means of sticks and straws, and that will be more profitable for the formation of their minds than the monotonous recitation of non-understandable rules, or of incomprehensible definitions.

They have, so far, only practised, by means of games, the two first rules of arithmetic ; it is only a short time since they were learning to write figures, or to form various letters, and now behold them plunged headlong, and you with them, right in the midst of Algebra.   If you mention this formidable word before them, do not fail to tell them that this science which is so useful and so wonderful, is comparatively speaking modern, and that it is François Viète [1] to whom the honour belongs of having been its inventor.

15. Calculations ; Measures ; Proportions.

We have seen from the beginning that what we are constantly doing is to calculate and to measure.   If we have before us a heap of grains of corn, and if we find upon counting them. that there are 157, this number, as we have previously noticed, would be useful in representing to us a collection of counters, of matches, of trees, of sheep, or in short anything.   If to determine length we have put sticks that are all alike one after another, and if we find 157 will measure this length, we say that it is the same length as 157 sticks.   In all these various cases we should never be able to value anything if we had not before us the idea of a grain of corn, a counter, a tree, a sheep, or a stick.

Number has no significance except by the comparison that it brings about with the single object (grain of corn, counter, etc.), without which it would be impossible to make it, and this single object is called "unity."   This comparison is what we term a proportion, and this idea of proportion leads us on to say that a number is simply a proportion of the number to unity.

It is all the more necessary to fix this firmly in the mind of the pupil, because unity is not always the same. For instance, having formed bundles of sticks, let us take a heap and count them ; we find there are 7 ; seven is

---

[1] Viète, French mathematician, born at Fontenay-le-Compte (1540—1603).

the proportion of our collection of sticks to one bundle,
which is unity. Now, let us scatter our sticks by
unfastening the bundles, and let us count ; it is the stick
which now becomes unity, and we can count seventy of
them ; this number will be the proportion of the collection
to one stick.

Similarly, let us take three faggots of sticks ; if we count
by bundles, we shall find thirty bundles ; but if by sticks,
three hundred.

Three will be the proportion of the whole heap of sticks
to a faggot ; thirty the proportion of the same heap to
a bundle ; three hundred the proportion to one stick.

We might give innumerable instances of such examples,
varying them indefinitely, in such a manner as to make
this idea of proportion perfectly familiar to the pupil.
This is the root of all calculation and all measurement,
but, by some strange hallucination, in academical teaching it is put at the end of arithmetic. It is not
possible to count two beans without having this idea
of the proportion of two to one ; nor to measure a length
of three yards without comparing the length with that
of one yard (proportion of three to one), and so on.

At this stage it will be desirable to *show* the pupil,
without any theoretical explanation, without any definition, without any appeal to his memory, the commonest
measures, weights and coins which we find ready to our
hand, yards, quarts, ounces, cents, etc.

We will give him exercises in making use of them,
accustoming himself to them to measure and to count,
and the idea of proportion will insensibly grow in his
mind, will associate itself indissolubly with that of number,
which is essential for the day in the future when he will
pass from play to work. And this work can become not
only interesting but amusing, instead of being a wearisome
task, if not a torture.

16. The Multiplication Table.

We are now going to learn how to make a little table which will be most useful to us because of what follows, and will be a very good exercise, even for its own sake. Under the form as represented below, this table is generally called the table of Pythagoras,[1] which may have been invented by this wonderful man, although it is not by any means certain ; however, even its reputed origin proves to us that it is not anything new.

To form the multiplication table we begin by writing on a sheet of squared paper the first nine numbers in the nine squares which follow each other :—

$$1 \quad 2 \quad 3 \quad 4 \quad 5 \quad 6 \quad 7 \quad 8 \quad 9.$$

Then, taking the first figure, 1, we add it to itself, which makes 2, which we write below; then 1 to 2, which makes 3; and so on, which gives the first column of Fig. 11.

We will do the same to fill the other columns, but the important thing is to write the results only and nothing else. For example, for the column which begins with 7, we would say, "7 and 7, 14 ; and 7, 21 ; and 7, 28 ; and 7, 35 ; and 7, 42 ; and 7, 49 ; and 7, 56 ; and 7, 63." And we write in succession 14, 21, 28, . . . 63 in the column beginning with 7.

1	2	3	4	5	6	7	8	9
2	4	6	8	10	12	14	16	18
3	6	9	12	15	18	21	24	27
4	8	12	16	20	24	28	32	36
5	10	15	20	25	30	35	40	45
6	12	18	24	30	36	42	48	54
7	14	21	28	35	42	49	56	63
8	16	24	32	40	48	56	64	72
9	18	27	36	45	54	63	72	81

FIG. 11.

It will be sufficient, we can see, for the pupil who knows his addition table, to be able to form this table very quickly. When he has completed it, we see that the rows and the columns are all alike. Thus the row which begins with 3 contains, like the column beginning with 3, the numbers 3, 6, 9, . . . 27.

[1] Pythagoras, Greek philosopher, born at Samos, 6th century B.C.

M.   D

It is absolutely necessary to plant this table firmly in our minds. But the proper way to arrive at this is not to attempt to learn it. This is accomplished by making it, verifying it, by carefully examining it, making use of it as we will show later on. If it does not come readily to the mind, the child must re-construct it—not a very

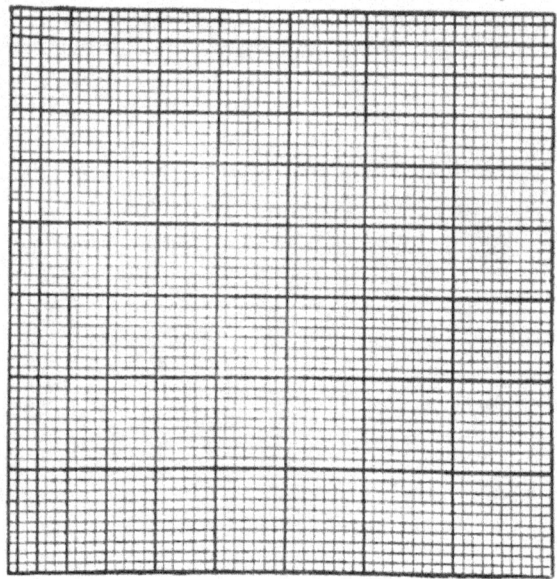

FIG. 12.

long business; and then he will finish by seeing it with his eyes shut.

We could make out the table beyond 9, but this is not advisable; because if we go on, for instance, up to 20 or 25, it will take much longer to do, and it is not necessary for the child to remember the table written out so far, although, of course, it would be useful.

He will notice certain peculiarities about this table. Thus in the column (or the row) beginning with 5, the

figures of the units are alternately 5 and 0 ; in the column
(or the row) beginning with 9, the figures of the units
8, 7, 6, . . . always diminish by 1, and those of the tens
1, 2, 3, . . . increase by 1. The explanation of this will
not be difficult to find.

It is worthy of notice that we can make a multiplication
table without using a single figure ; to do this, it is only
necessary to have a squared paper in sufficiently small
divisions (Fig. 12). The table shown in this figure is
made out as far as 10. We proceed as follows : mark
divisions out successively, on a horizontal line, 1, 2, 3,
. . . 10, and mark the points of division. Then, on a
vertical line, taking the same starting point, do the
same thing ; mark off the points of division by means
of strong lines, and we obtain large divisions ; and each
of these divisions contains a number of little ones which
will be precisely the same as is contained in our table of
figures. The explanation of this is quite simple ; for
our table in Fig. 12 only represents by means of a series
of lines what in Fig. 11 is done by calculation.

17. Products.

If I take a heap of 7 sticks, and I form three such heaps,
I can ask myself, "How many sticks will there be in all? "
That is called multiplying 7 by 3. The result obtained
by such multiplication is the product of the two numbers
7 and 3; 7 is the multiplicand, and 3 the multiplicator.
If, instead of heaping up the sticks, we leave the three
heaps separate, we see that (taking one heap as representing a unit) the number showing the product will be
3 ; or that the proportion of the product will be 3 ; which
number also gives us the proportion of 3 to 1.

So we can say equally well that to multiply 7 by 3 is
to repeat 7, 3 times ; or to find a number of which the
proportion to 7 will be the same as that of 3 to 1.

To multiply a number (the multiplicand) by another

(the multiplicator) is to find another number (the product), which may be formed by repeating the multiplicand as many times as there are units in the multiplicator; this product, in short, bears the same proportion to the multiplicand which the multiplicator does to unity.

These are not formulæ which the child must learn; they are ideas which he must insensibly assimilate by our aid ; for the former present an appearance of difficulty, while the latter can be easily grasped by the pupil, especially if we take the trouble of expressing them by means of grains of corn, of sticks, or of divisions on squared paper.

It is quite certain that the child will jump immediately to the idea that, to find a product, he has nothing to do but make an addition, that the product of 7 by 3 is $7 + 7 + 7$ in the same way that 3 is $1 + 1 + 1$. And as the table (Fig. 11) has been made in this way, it gives us the desired product by taking the column that begins with 7, the line which begins with 3 and looking for the meeting point, where we read 21.

Explain at this point that the multiplication sign is $\times$, and that therefore the phrase " the product of 7 multiplied by 3 is 21 " is expressed thus : $7 \times 3 = 21$.

7×3 is often written 7.3 ; instead of these two we can have any two numbers whatever represented by the letters a, b. Their product can be expressed by $a \times b$, or $a.b$, or simply ab ; when we write $ab = p$, we mean that a multiplied by b is p.

It is well to know also that we can consider products such as $a \times b \times c \times d$, or simply $abcd$, for instance ; this means that we multiply a by b, then the resulting product by c, then the new product by d ; a, b, c, d, are called the *factors* of the product $abcd$; we can thus have the product of any number of factors.

As regards the practice of multiplication, we must first notice that the table gives us the answers when the multiplicand and the multiplicator are each less than ten.

It will be easy afterwards to show how we can multiply a number by 10, 100, 1,000.

The employment of numerous examples conforming to all the rules given in all the arithmetic books might be useful, provided that it is not accompanied by any theory. I cannot sufficiently impress upon you the importance of giving preference to the Mohammedan method, which is almost as quick, much easier to understand and carry out, and not sufficiently known in teaching although approved by several writers.

We are going to take the very simple instance of 9,347 × 258. The multiplicand has 4 figures, and the multiplicator has 3; let us mark out on squared paper 3 rows of 4 divisions each; on top of this figure let us write the figures of the multiplicand 9, 3, 4, 7, *from left to right*; at the left and working *from the bottom to the top*, those of the multiplicator 2, 5, 8; having traced the dotted lines of the figure,

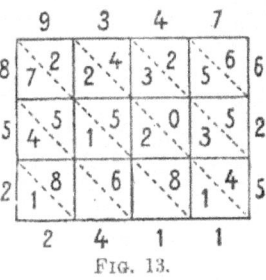

FIG. 13.

let us now put in each division the product of the two corresponding numbers, as though we were making a multiplication table, but always placing the figure of the tens of the product *below*, and that of the units *above* the dotted line; finally, we add up, taking for the direction of the columns the dotted lines; thus we find the product 2,411,526. The great advantage of this method is that it does not necessitate any partial multiplication nor the observance of any special order. As all the divisions are filled we are certain to forget nothing.

With the same example, and the same method, we indicate a slightly different arrangement which does not compel us to add up obliquely, and which in this sense is perhaps most convenient (Fig. 14).

As regards the justification of the Mohammedan method, it is sufficiently evident to everyone acquainted with the theory of multiplication, although at present unnecessary for the child. If he is of an enquiring disposition he will probably discover it for himself. What is of real importance is that he shall be able to calculate correctly, and that he *will be interested in doing so.* From the moment when fatigue or boredom overtakes him it is absolutely necessary to go on to something else.

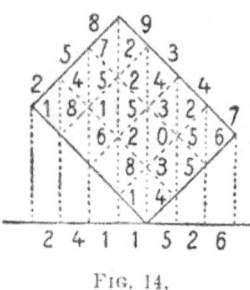

Fig. 14.

However, we will not abandon what relates to multiplication without recalling that a product

$$a \times a \times a \times \ldots \times a,$$

of which all the factors are equal, is called a *power* of a; that such a product is written a^n, n being the number of the factors, and that we call it the nth power of a: that the second power is called *square*, and the 3rd *cube* (we shall soon see the reason for this). The number n is called the *index*.

$$2 \times 2 \times 2 \times 2 = 2^4 = 16; \text{ the index is 4.}$$

The cube of 5 is

$$5 \times 5 \times 5 = 5^3 = 125; \text{ the index is 3.}$$

The square of 7 is

$$7 \times 7 = 7^2 = 49; \text{ the index is 2.}$$

18. Curious Operations.

There are a number of results from operations which strike us because of their peculiarities. Their great merit consists in arousing the child's curiosity, and thus giving him a taste for calculation.

Subjoined are various examples which will be sufficient for our purpose.

I. Ask the child, after giving him a sealed envelope, to write a number of 3 figures ; let him choose one to suit his own fancy. Suppose he writes 713 ; then write it the reverse way, which makes it 317 ; next, subtract one from the other, which gives the result 396 [1] ; reverse this number, which then reads 693 ; finally add these last two, which will be 1089. At this point ask him to open the envelope ; he will find in it a paper on which you have written beforehand 1089. It is remarkable that any other number of 3 figures would have led to the same result, provided that the first and third figures are different.

II. If the child works out $12 \times 9 + 3$, $123 \times 9 + 4$, and so on, as far as $123456789 \times 9 + 10$, each result will contain no other figure than 1 written a varying number of times, $12 \times 9 + 3$ giving us as result 111, and $123 \times 9 + 4$ equalling 1111, and so on.

On the contrary, working out $9 \times 9 + 7$, $98 \times 9 + 6$, and so on till we reach $9876543 \times 9 + 1$, we shall have results which will contain no figure but 8.

III. The product of 123456789×9 will be a number made up of 1's. Taking the same multiplicand and multiplying it by

18 ($= 9 \times 2$), 27 ($= 9 \times 3$) . . . to 81 ($= 9 \times 9$),

we shall find some curious products, all made up of the same figure repeated.

IV. Take the number 142857 ; if we multiply it successively by 2, 3, 4, 5, 6 our products will be

285714, 428571, 571428, 714285, 857142.

It will be seen that each product contains the same figures as the multiplicand.

---

[1] This difference ought always to have three figures. If there are only two, we must put a nought in the hundreds place. For example, 716 and 617 will give us for difference 099; according to this rule, adding 099 and 990 we make 1089.

Multiplying it by 7, we have 999999, and if we cut it in two, making 142 and 857, the sum of these numbers will be 999. The same result is obtained by cutting in two any of the five products written above.

V. Complete these examples by showing the easy method of multiplying by 9, by 99, 999. This should always be done by multiplying by 10, 100, 1000, etc., and then subtracting the multiplicand. If the numbers are not too big this can quite soon be done mentally.

19. Prime Numbers.

Running our eyes over a multiplication table, we notice that it includes certain numbers up to the limit to which it extends, but not all these numbers. In other words, there are numbers which are products, and others which are not ; the latter are called the prime numbers, the others are called composite numbers.

For instance, 2, 3, 5, 7, 29, 71 are prime numbers ; 4, 6, 9, 87, 91, are composite numbers, because $4 = 2 \times 2$, $6 = 2 \times 3, 9 = 3 \times 3, 87 = 3 \times 29, 91 = 7 \times 13$. However far we advance in the series of numbers we always meet these prime and composite numbers. This distinction is fundamental ; and yet, despite the labours of our most learned men, we know very little about prime numbers. We are incapable, if the number under consideration is rather large, of saying whether it is prime or not, unless we give ourselves up to a groping or tentative method which requires very long and laborious calculations. This shows us how very little progress science has made on these questions which appear quite simple, and how modest we ought to be when we compare the slight extent of our knowledge with the immensity of the things about which we are ignorant.

Still, from antiquity we have possessed a method of finding the prime numbers to a limit as distant as we wish. This method consists in writing out the complete list of these numbers, then cancelling those which are products of 2, of 3, of 5, etc. We intend applying it here to the first 150 numbers. Write them, suppressing 1, which is useless.

2 3 ~~4~~ 5 ~~6~~ 7 ~~8~~ ~~9~~ ~~10~~ 11 ~~12~~ 13
~~14~~ ~~15~~ ~~16~~ 17 ~~18~~ 19 ~~20~~ ~~21~~ ~~22~~ 23 ~~24~~ ~~25~~
~~26~~ ~~27~~ ~~28~~ 29 ~~30~~ 31 ~~32~~ ~~33~~ ~~34~~ ~~35~~ ~~36~~ 37
~~38~~ ~~39~~ ~~40~~ 41 ~~42~~ 43 ~~44~~ ~~45~~ ~~46~~ 47 ~~48~~ ~~49~~
~~50~~ ~~51~~ ~~52~~ 53 ~~54~~ ~~55~~ ~~56~~ ~~57~~ ~~58~~ 59 ~~60~~ 61
~~62~~ ~~63~~ ~~64~~ ~~65~~ ~~66~~ 67 ~~68~~ ~~69~~ ~~70~~ 71 ~~72~~ 73
~~74~~ ~~75~~ ~~76~~ ~~77~~ ~~78~~ 79 ~~80~~ ~~81~~ ~~82~~ 83 ~~84~~ ~~85~~
~~86~~ ~~87~~ ~~88~~ 89 ~~90~~ ~~91~~ ~~92~~ ~~93~~ ~~94~~ ~~95~~ ~~96~~ 97
~~98~~ ~~99~~ ~~100~~ 101 ~~102~~ 103 ~~104~~ ~~105~~ ~~106~~ 107 ~~108~~ 109
~~110~~ ~~111~~ ~~112~~ 113 ~~114~~ ~~115~~ ~~116~~ ~~117~~ ~~118~~ ~~119~~ ~~120~~ ~~121~~
~~122~~ ~~123~~ ~~124~~ ~~125~~ ~~126~~ 127 ~~128~~ ~~129~~ ~~130~~ 131 ~~132~~ ~~133~~
~~134~~ ~~135~~ ~~136~~ 137 ~~138~~ 139 ~~140~~ ~~141~~ ~~142~~ ~~143~~ ~~144~~ ~~145~~
~~146~~ ~~147~~ ~~148~~ 149 ~~150~~

Starting from 2, and going on by 2's, according to the list, we find the products of 2, which are 4, 6, . . ., that is to say, even numbers; they are not, therefore, prime numbers, and we cancel them at one blow as far as 150.

Let us begin this time with 3 and continue by 3's, cancelling the numbers which we find to be products of multiplication by 3, unless they have been cancelled already.

The first number uncancelled after 3 is 5; starting therefore from 5, and moving on by 5's, we proceed in the same way. Doing the same with 7 and 11, we find that only the following numbers remain uncancelled :—

2 3 5 7 11 13 17 19 23 29 31 37
41 43 47 53 59 61 67 71 73 79 83 89
97 101 103 107 109 113 127 131 137 139 149 .

which gives us the list of prime numbers up to 150.

This ingenious process is known as the sieve of Eratosthenes[1] from the name of the inventor.

20. Quotients.

Some small heaps of counters, each containing 7, have been formed into one large one, which contains 56. We want to find out how many little heaps we have used to make the large one.

What we are now going to do to discover this is called a *division*. We can also say that, knowing the product of two factors, 56, and one of the factors, 7, we wish to find the other.

The product given, *i.e.*, 56, is called the *dividend*; the factor given, *i.e.*, 7, is called the *divisor*, and the result we are seeking is the *quotient*.

We really could find the quotient by subtraction, that is to say, by taking the divisor from the dividend, then the divisor from the remainder obtained, and so on, until there is nothing left, being careful to count how many subtractions we have made. Thus, successively taking away 7 from 56, we have the numbers 49, 42, 35, 28, 21, 14, 7, and we see that we have needed to subtract 8 times to exhaust our dividend 56. The quotient we want is thus 8; naturally, our knowledge of the multiplication table would show us this immediately.

The method of successive subtractions would be impracticable with rather large numbers, because of its length. The classical rule that is adopted in doing division is nothing else than a means of counting very much faster subtractions done all at once.

To accustom children to be able to divide with ease, we must begin by always making them form into a little table the products of the divisor by 2, 3, . . . 9. This does away with the hesitation that so hampers the beginner in all his work.

---

[1] Eratosthenes, Alexandrian scholar, born at Cyrene (276—193 B.C.).

We will show him how this is done by the example given below of the division of 643734 by 273. The products of 273 are

1 × 273 =	273	6 × 273 =	1638
2	= 546	7	= 1911
3	= 819	8	= 2184
4	= 1092	9	= 2457
5	= 1365		

Let us work out the problem in the ordinary way:—

```
643734 (273
546      2358
─────
977
819
─────
1583
1365
─────
2184
2184
─────
   0
```

In the dividend there are 643 thousand ; as our little table shows us, 643 contains the divisor (273) twice; therefore, taking away 546 thousand from the dividend, we make 2 thousand subtractions at once. We have left 97734, which contains 977 hundreds ; 977 contains the divisor 3 times, so taking away 819 hundreds we have made 300 subtractions at once. Now our remainder is 15834, containing 1583 tens ; 1583 contains the divisor 5 times, and taking away 1365 tens, we make this time 50 subtractions. Finally, there remains 2184, which contains the divisor exactly 8 times ; therefore, by subtracting 8 times more we shall have taken everything from the dividend, and the quotient will be 2358, the total number of subtractions.

The child should form the habit of dividing in this way

without it being necessary to explain in great detail what we have just expressed at length.

The above division could be done because the divisor and dividend had been picked, but if we take two numbers at random, it is hardly likely such division will be possible. We will not be able to take away the divisor a certain number of times, so that nothing is left. However, if we work on precisely the same lines, taking the divisor from the dividend *as many times as we can*, we shall come at last to a point when the dividend will be smaller than the divisor. This number is termed the *remainder* of the division.

As a very simple example, let us suppose that we are going to divide 220 by 12. We shall discover this to be impossible, for when 12 has been taken 18 times from 220, 4 will remain ; it follows that $220 - 4$, or 216, would be divisible by 12, and the quotient would be 18. These impossible divisions, when we get a remainder, can easily be turned into divisions which are possible, by simply replacing the dividend by this dividend with the remainder taken from it.

Do not, however, lay much stress on this operation of division, except from the point of view of the practice of calculation. Theories are interesting, but will be more useful at a later period. They are hardly suitable for the introductory period.

It is useful to know that division is shown by a sign $\div$ or —.[1] Thus $56 \div 7$ or $\frac{56}{7}$ expresses the quotient of 56 by 7. We can write $56 \div 7 = \frac{56}{7} = 8$. In general, $\frac{a}{b} = q$ means that the quotient of the division of a by b is the number q.

[1] As the minus sign and the division sign (2nd method) are similar, it would be well to point out that the minus sign is placed between two numbers, one following the other, whereas the division sign is placed between two numbers, one placed above the other. [Tr.]

21. The Divided Cake; Fractions.

Suppose five people wish to share a round cake equally. It will have to be cut into five perfectly equal pieces (Fig. 15) by incisions starting from the middle, and one of these pieces, as AOB, will be the share of each one of the five. This part is called a fifth of the cake, and AOB will be represented by $\frac{1}{5}$ of a cake.

Of the five people, two may be absent, but their shares are to be reserved for them, so the parts AOB, BOC, both exactly alike, are put on one side for them. These two pieces taken together are called two-fifths of a cake, and they are shown by the figures $\frac{2}{5}$. All these numbers, such as $\frac{2}{5}$, are called *fractions*; 2 and 5 are the two *terms*; 2, which is above the line, is the *numerator*, indicating the number of

FIG. 15.

pieces; 5, which is below, is the *denominator*, showing into how many equal pieces we have divided the cake.

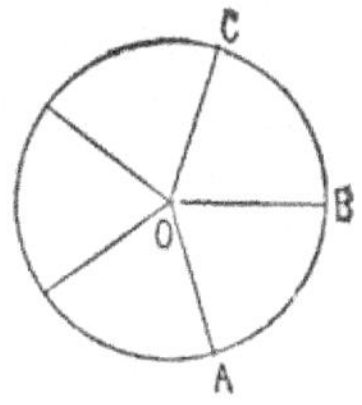

If we had taken $\frac{5}{5}$ of the cake, we can see that would have been the whole cake; and if the cake had been divided into any number whatever of equal parts, by taking afterwards the same number of parts the cake would be again entire, so that the fraction $\frac{a}{a}$ of which the numerator and denominator are alike is always equal to 1.

But suppose 10 people, and not five, wanted to divide the cake; then we should have had to divide the cake into 10 equal parts, and the tenths could be obtained by taking fifths, such as AOB, and cutting each of them into two equal parts. Then $\frac{2}{10} = \frac{1}{5}$; and it is no harder to see that two fractions are equal if we can pass from one to another by multiplying the two terms by the same number. This fundamental principle of the whole theory of fractions is proved in this manner by a sort of intuitive evidence, by means of concrete objects, and we need ask for no demonstration.

Suppose now, there are 17 cakes, all alike, and that 5 people want to divide them equally. There are two ways of doing this. We can divide each cake into fifths, and each person can take a fifth of each cake, in all $\frac{17}{5}$, or, secondly, we can give each one an equal number of cakes, which of course, for five people will be 3, as 5 is contained in 17 three times. Then we shall be left with 2 cakes to divide. Dividing them into fifths, each of the five persons will take $\frac{2}{5}$; by this means we see that $\frac{17}{5} = 3 + \frac{2}{5}$.

It is easy by this means to initiate the child into all the ordinary fractional calculations, on which it seems unnecessary to dilate longer. But we can only succeed by always using concrete objects, such as cakes, apples, oranges, divided lengths, etc. Then he will seize the idea wonderfully well, that these new arithmetical expressions are numbers, and that they express proportions.

We ought at this stage to make it clear that these numbers can only be used in relation to those things which, by their very nature, are divisible, such as those we have indicated above; if, for example, a question involving the consideration of a certain number of persons was placed before us, the application of fractional numbers would be absurd, and, as the result would show, impossible. It is to be regretted that this is not more frequently mentioned to the pupil.

In other words, calculations adapt themselves to suitable things, which are very numerous, but not

universal.   And this further maxim must always be borne
in mind by the child, that in arithmetic there is a necessity
for the exercise of his own reflection and common sense,
without which a mere dexterity in computation will be
useless.

For instance, here is a problem, particularly mentioned
by Edward Lucas,[1] which will serve as a useful exercise
on this maxim.   A tailor has a piece of stuff 16 metres
in length, from which he cuts off 2 metres each day ;
in how many days will he have cut up the whole piece ?
A want of thought, joined to the habit of mechanical
calculation, leads to the answer 8, instead of the number
7, which an exercise of common sense would indicate.

Questions involving fractions should be varied, not
too complicated, and borrowed from effective concrete
subjects.   It would be wise to complete them by drawing
attention to decimal fractions, to the manner in which
they may be written, and to the methods of calculation
connected with them.

Many good treatises on arithmetic will furnish the
necessary instances in this respect.   I content myself
with insisting on the usefulness, above all, of measures
of length, and instances taken from the counting of
money.

Finally, we must notice that if the sign of division and
notation of fractions are alike, it is not by chance, and
does not give rise to confusion; $\frac{15}{3}$, for instance, expresses
not only the quotient of the division of 15 by 3 but the
fraction $\frac{15}{3}$.   This can be seen with concrete objects, and
we only need mention it.

The properties and calculation of fractions can also be

---

[1] Edward Lucas, French mathematician, born at Amiens (1842—1891).
He was, perhaps, the man who, in his day, understood better than any
other the science of numbers.   His merit has been singularly unrecog-
nised, and this contributed to his premature death.

shown in a very simple fashion, by making use of squared paper.

Judgment can be passed upon it by the remarks which follow. They only presented themselves to my mind after the publication of the second edition. I will endeavour to explain them now, as briefly as is consistent with a sufficient clearness of expression. I address myself to teachers, and, provided they have understood what I have said, they will have no trouble in using the means proposed, under whatever form seems good to them— that is, of course, if they agree with the principle of my point of view.

Suppose, then, we represent a concrete unit of any kind

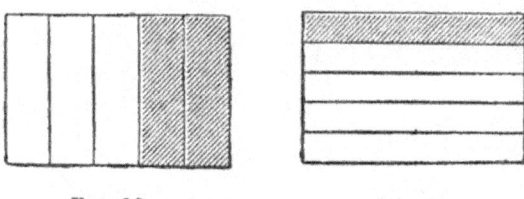

FIG. 16.   FIG. 17.

(provided that, by its nature, it is divisible) by means of a rectangle (Figs. 16, 17).

If we divide the length of this rectangle into 5 equal parts, we can cut it into 5 portions, into 5 vertical bands all alike. Each of them will be *a fifth* of the unit (Fig. 16).

If we divide the height of the rectangle into 5 equal parts, we can also cut it into 5 horizontal bands, each being alike, and each of them, also, a *fifth* of the unit (Fig. 17).

Again, divide the length (Fig. 18) into 3 equal parts, and the height into 4 such parts, the unit can be cut by lines passing through the points of division into 12, or 3 × 4 little rectangles, all alike, and each being a *twelfth* of the unit.

Taking any number whatever of these bands or rect-

angles, we have what is called a *fraction*.   If the number
of bands or little rectangles is less than those which make
up the unit we have a fraction properly so called, or a
proper fraction.   If greater, we have a fractional expression or an *improper fraction*.   Therefore a proper fraction
is less than 1, and an improper fraction is greater than 1.
If, by taking just the same number of bands or rectangles
as there was in the unit, we re-form the unit, such a
fraction, consequently, is equal to 1.

When we use the word " fraction " the word means,
as a general rule, a proper fraction.   By means of shading,
we can express any fraction whatever, leaving white those

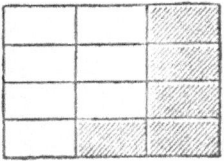

FIG. 18.   FIG. 19.

parts which we take from the unit, and making dark
the bands or rectangles which are left.   Thus in Fig. 16,
we see the fraction three-fifths ; in Fig. 17, four-fifths ;
in Fig. 18, seven-twelfths.

The number of white rectangles (3, 4, 7 in the three
examples) is called the numerator ; the whole number of
rectangles into which the unit is divided (5, 5, 12) is
called the denominator ; and the three fractions are
written thus $\frac{3}{5}, \frac{4}{5}, \frac{7}{12}$.   If we need to represent an improper
fraction, we should have no rectangles shaded, and the
symbol would be greater than the unit, the numerator
larger than the denominator.   If the numerator is the
same as the denominator, we have the unit itself.

Fundamental Principle.—The value of a fraction is in

M,   E

no way changed by multiplying the numerator and denominator by the same number.

Let us take the fraction $\frac{3}{4}$ (Fig. 19). I wish to show that it is equal to $\frac{15}{20}$ or $\frac{3 \times 5}{4 \times 5}$. The fraction $\frac{3}{4}$ was represented by vertical bands. I divide the height into 5 equal parts and I picture the rectangular unit cut up by horizontal lines passing through the points of division. It is thus divided into little rectangles all alike; and we shall see that we have now 4×5 or 20; then take the fraction $\frac{3}{4}$; it contains 3×5 or 15 little rectangles; it has not changed; its denominator and its numerator have each been multiplied by 5; therefore $\frac{3}{4} = \frac{3 \times 5}{4 \times 5} = \frac{15}{20}$.

Two fractions can be reduced graphically to the same denominator, either by working on the above principle, or dealing directly with the figure.

We will take, for example, $\frac{1}{4}$ and $\frac{2}{3}$, the first fraction represented by a vertical band, and the other by two horizontal bands (Fig. 20).

Dividing the first rectangle into three horizontal bands exactly similar, the second into four vertical bands, we see that our two fractions read $\frac{3}{12}$ and $\frac{8}{12}$.

Addition and subtraction will then be readily explained, taking for illustration these concrete expressions.

For multiplication, the definition itself tells us that, to multiply $\frac{2}{5}$ by $\frac{3}{4}$, we must take $\frac{3}{4}$ of $\frac{2}{5}$.

Take (Fig. 21) the fraction $\frac{2}{5}$ represented in ABCD by two vertical bands. Dividing the height AD into four equal parts, and drawing horizontal lines, we have divided $\frac{2}{5}$ into four equal parts; take three of these, and let us darken the remainder by horizontal shading.

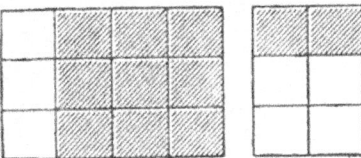

FIG. 20.

The white part is the product $\left(\frac{3}{4} \text{ of } \frac{2}{5}\right)$: it contains 2×3 or 6 little rectangles; and the unit contains 5×4, or 20. Therefore

$$\frac{2}{5} \times \frac{3}{4} = \frac{2 \times 3}{5 \times 4} = \frac{6}{20}.$$

It is easy by such means to formulate in the child's mind the idea of *proportion* (the proportion of a to b being the number which gives the measure of a when we take b for the unit), to show the identity of this proportion with the fraction $\frac{a}{b}$, to establish that $\frac{a + m}{b + m}$ approximates indefinitely to 1 when we give to m increasing integral values, to make it understood that a fraction is the

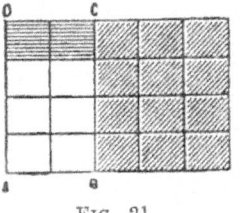

FIG. 21.

quotient of the numerator by the denominator, to explain clearly the fundamental principles of proportions, etc.

Whether with squared paper, or with little squares or rectangles of wood, white on one side and black on the opposite one, all these operations can be carried out in the pupil's sight. When using materials such as have been mentioned, the work is at once amusing and instructive; it rivets the child's attention, it fixes in his mind the essential truths, without which he may be obliged to force his memory; he *sees* these truths, he makes them, as it were, with his own hands; they are no longer for him obscure phrases repeated without any meaning attached to them, but tangible realities. Experience has shown that these methods are most efficacious from a scholastic point of view; it is most desirable that their use should be more and more extended.

22. We Start Geometry.

We have already seen what is a straight line. It is the most simple of all the geometrical figures. We can try and extend our knowledge a little in this respect.

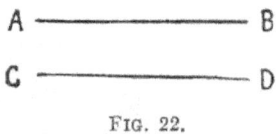

FIG. 22.

Let us begin, for instance, by forming an idea of a plane by looking at the surface of a smooth piece of water, that of a good looking-glass, of a ceiling, a floor, or a door. A slate, a sheet of paper stretched on a polished board, give us also the idea of a flat surface, and we feel that, like the straight line, the plane may, in thought, be prolonged as far as we like, indefinitely. On a flat surface we can lay a ruler in any direction. On a sheet of paper we can draw as many lines as we wish.

If we draw two only, they can be (Fig. 22) parallel, as AB, CD. On ruled paper it can be plainly seen that the

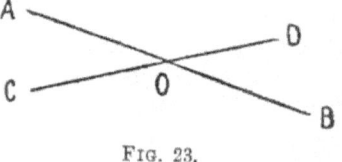

FIG. 23.

lines on it are all parallel, and that two parallel lines never meet. If, on the contrary (Fig. 23), the two straight lines AB, CD, meet at the point O, the figures

AOC, COB, BOD, DOA are angles. Two angles are equal when we can place one over the other. The angles AOC, BOD, for example, are equal; it is the same with COB, DOA.

When two straight lines (Fig. 24) cut each other in such a way that the angles DOA, AOC are equal, the four angles round O are all equal; we call them then *right angles*, and the figure formed by the two lines is that of a cross. On squared paper, we can see right angles everywhere where two lines meet. When two straight lines form right angles in this way we say that they are *perpendicular* to one another.

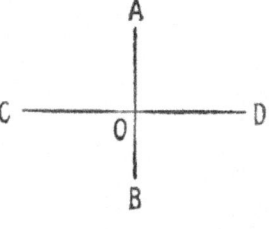

FIG. 24.

An angle less than a right angle, like AOC, in Fig. 23, is called an *acute* angle; if greater than a right angle, like COB, it is an *obtuse* angle. A plumb-line shows a straight line which is called *vertical*. A straight line perpendicular to a vertical line is called *horizontal*. All straight lines drawn on the surface of smooth water would be horizontal straight lines, and this surface is in itself a plane, which we call *horizontal*. On a piece of ruled paper laid before us, the lines which run from left

FIG. 25.

to right are called horizontal, and the others vertical, because we suppose that the paper is lifted up and laid against a wall.

Let us imagine that on a sheet of paper we have drawn three straight lines. These may be considered in several ways. The three straight lines (Fig. 25) can be parallel. In the second place (Fig. 26) two, AB, CD, might be parallel, and the third, EF, might intersect them at E, F.

This third line is called a *secant*. In this figure all the angles marked 1 are equal to each other; the angles marked 2 are also equal to each other; and the sum of an angle 1 and of an angle 2 is equal to two right angles. It might happen (Fig. 27) that our three straight lines

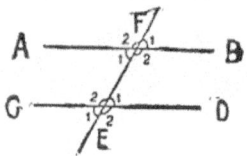 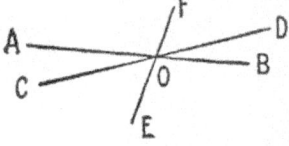

FIG. 26.　　　　　　　　FIG. 27.

passed through the same point O; we would say then that they were *concurrent*.

Finally (Fig. 28), f none of the preceding circumstances occur, the three straight lines will cross each other twice at the three points A, B, C, and will limit a portion of the plane ABC, which we can consider apart (Fig. 29), and which is called a *triangle*. The points A, B, C are called the *apexes*, and the segments, AB, BC, CA the *sides*, of the triangle. We say that the angles A, B, C marked on the figure are the *angles* of the triangle.

One of the angles of the triangle can be a right angle; we say then that the triangle is *right-angled*. It may also (Fig. 31) have an obtuse angle; we say then that

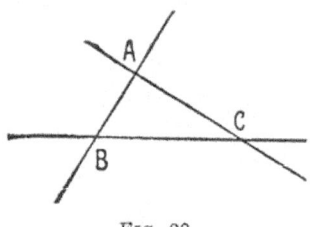

FIG. 28.

the triangle is *obtuse-angled*.

If a triangle like those of Fig. 32 has two sides AB, AC which are equal, the triangle is termed an *isosceles* triangle. The angles B and C are then equal.

If a triangle has its three sides equal, it is *equilateral*. Its three angles are then equal (Fig. 33).

In a triangle ABC (Fig. 34), we can choose any side, BC, and call it the *base*. If we then draw from the point A a straight line perpendicular to BC, and which

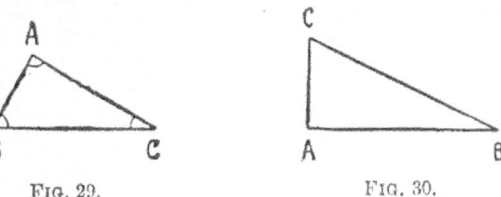

FIG. 29.   FIG. 30.

meets BC at A', we say that AA' is the *height* or *altitude* of the triangle.

This simple figure, the triangle, has innumerable

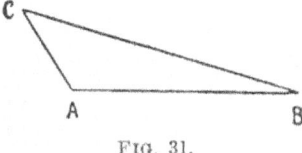

FIG. 31.

properties; we will examine some of them later. At the moment we must not deceive ourselves: we are learning nothing at all; we are simply looking at the

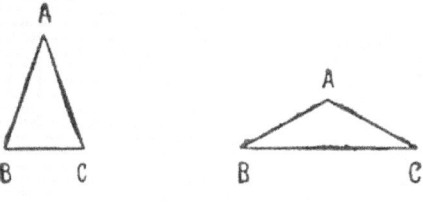

FIG. 32.

figures and getting to know their names. That is something useful, at any rate.

When a part of a plane (Fig. 35) is limited by several straight lines, or rather by several segments of straight lines, this figure is called a *polygon*. The segments AB,

BC, . . . HA are the *sides*, the points A, B, . . . H, the corners or *apexes*, the angles marked A, B, . . . H the *angles* of the polygon.

A polygon like that of Fig. 36 is said to have *re-entering*

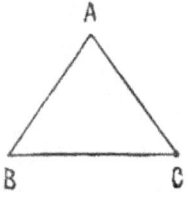

FIG. 33.

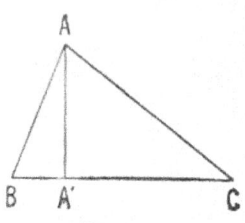

FIG. 34.

angles. When there is no re-entering angle, as in Fig. 35, the polygon is *convex*. Generally speaking, we shall only deal with convex polygons. A straight line like AD (Fig. 35), which joins two corners of a polygon, and which is not a side, is called a *diagonal*.

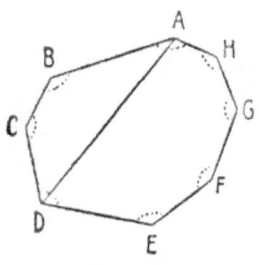

FIG. 35.

In a polygon, the number of the corners, the sides, and the angles are the same. Special names have been given to different polygons, according to the number of sides which they possess. To begin with, as we have already said, a polygon with three sides is a triangle. Then

A polygon of 4 sides is a quadrilateral

		5		pentagon
,,	,,	6	,,	hexagon
,,	,,	7	,,	heptagon
,,	,,	8	,,	octagon
,,	,,	10	,,	decagon
,,	,,	12	,,	dodecagon.

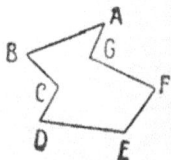

FIG. 36.

Thus, Fig. 35 represents a convex octagon, and Fig. 36 is a heptagon with re-entering angles. In a quadrilateral, the two sides AB, CD can be parallel (Fig. 37), the two others not being so; such quadrilaterals are

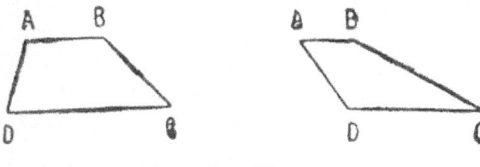

FIG. 37.

called *trapeziums.* The sides AB, CD are the *bases* of the trapezium.

If (Fig. 38) the sides AB, CD are parallel, and if the sides BC, AD are also parallel, the quadrilateral is a *parallelogram.* Then the sides AB and CD are equal, and so are the sides BC and AD. Also, the angles A, C are equal, and equally so the angles B, D.

If in a parallelogram the four sides are equal, it is a *rhombus* (Fig. 39).

If (Fig. 40) one of the angles is a right-angle, the three

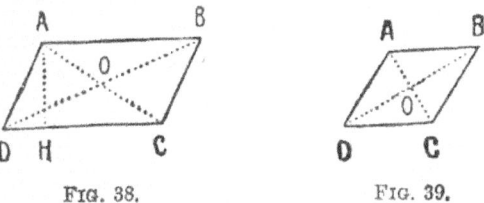

FIG. 38.   FIG. 39.

others are right-angles also, and the parallelogram is a *rectangle.*

If, finally (Fig. 41), a rectangle has all its sides equal, it is called a *square.*

In every quadrilateral there are two diagonals; in every parallelogram (Figs. 38, 39, 40, 41) these two diagonals AC, BD cut each other at a point O, which is the middle of each of them. In a rhombus (Fig. 39) the two diagonals are perpendicular to one another.

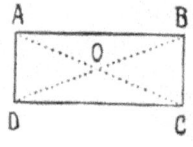

FIG. 40.

In a rectangle (Fig. 40) the two diagonals are equal. In a square (Fig. 41) the two diagonals are both equal and perpendicular. It will be noticed that a square is both a rhombus and a rectangle.

If in the parallelogram (Fig. 38) we take a side CD, which we call the *base*, and if we have a straight line AH perpendicular to CD (and also perpendicular to AB), this straight line, or rather this segment AH, is termed the *height* of the parallelogram.

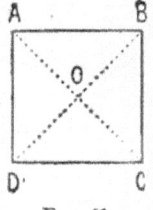

FIG. 41.

On squared paper, by following the squared lines, we can form as many rectangles and squares as we like. The various figures of which we have spoken, and others which we can imagine, should be constructed time after time by the pupil, with the help of a pencil, a ruler, a set square, and a measure to measure the lengths. Every line should be drawn with the greatest possible care. Afterwards he must accustom himself to draw them correctly without the help of any instrument. For this purpose it will be well for him, after having drawn his figure in pencil with his instruments, to draw it freehand afterwards in ink.

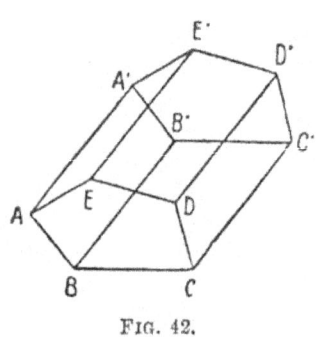

FIG. 42.

We are not saying anything yet about the use of the compass, and the protractor; we shall touch briefly upon this later on.

Moreover, do not let us lose sight of the fact that the child must never have left off drawing since he began to trace his first strokes.

When (Fig. 42) we have a polygon, ABCDE, on a horizontal plane, for instance, if we draw the straight lines AA', BB', CC', DD', EE', all parallel and equal to one another, outside the plane, the extremities, A', B', C', D', E', are the corners of another polygon like the first. The quadrilaterals AA', BB' ... are then parallelograms; the space which would be limited by all these parallelograms and by the two polygons is called a *prism*; the two polygons are the *bases*; the parallelograms are the

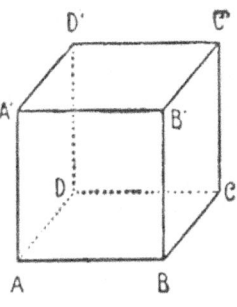

FIG. 43.

faces; the distance between the planes of the two bases is called the *height*. The straight lines AA', BB' ... are the *edges*.

If the edges are vertical (supposing the bases to be horizontal) the prism is *right-angled*.

If the bases are parallelograms, the prism is called a *parallelopiped*.

If, finally, the base is a square, and if the parallelopiped, being right-angled, has for its height the side of the base, the parallelopiped then takes the form of one of a set of dice, and is called a *cube* (Fig. 43).

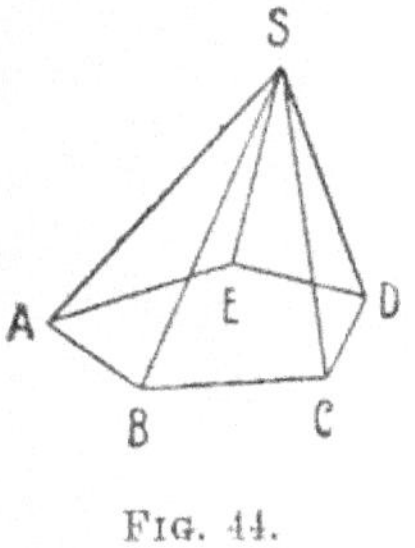

FIG. 44.

When (Fig. 44) we have a polygon, ABCDE, if we join all the corners with a point S outside the plane, the space which would be limited by the polygon and the triangles SAB, SBC, ... SEA, is called a *pyramid*; ABCDE is the *base*; the triangles SAB ... are the *faces*; SA, SB, ... are *edges*; S is the *apex*;

the distance from the apex to the plane of the base, which would be vertical if the base was horizontal, is the *height* of the pyramid.

With some small sticks and bits of wire it is easy to learn to construct little models giving a sufficiently exact idea of the figures we have just mentioned. We can also cut them with a knife out of a carrot, or a potato

Notice that the Figs. 42 and 44, in perspective, are made assuming all the edges to be visible (figures in little sticks), whilst in the cube of Fig. 43 we find three unseen edges AD, DC, DD' (indicated by dotted lines), which takes place if the cube is a solid body.

23. Areas.

The word " geometry " signifies, by its etymology, the measurement of the earth. That hardly answers to geometrical science such as we know it to-day, but it throws a light upon the origin of this science, which has arisen, like others, from the needs of the human race. From quite early times men have recognised the need for estimating the extent of pieces of land, and have sought the best means of arriving at it. These pieces of land being pretty nearly flat on the whole and more often than not limited by straight lines, it follows that, to acquaint ourselves with the extent of land, we must determine and estimate the extent of the various polygons described in the preceding chapter.

But to measure anything, no matter what, we must have a unit. We know how to measure lengths, taking for a unit a metre, or a match, or the side of a division of squared paper. To measure length we must have a unit of length. To measure a flat expanse, which is called an *area*, we must start with a unit which is in itself an area.

Invariably, the unit of length having been chosen the area unit will be the area of the square having for its side this unit of length.

If, to measure a length, it suffices to lay out the chosen unit end to end, and to count the number of times that we have thus laid it out, it is easy to understand that such a proceeding is practically impossible when dealing with an area; the squared unit must be laid out so that the area will be covered, which cannot be done.

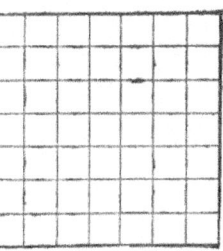

On the other hand, for the figures described further back, there are very simple ways of determining their areas.

To begin with, let us take a square. We will take squared paper, of which we suppose that each division is the unit of length.

FIG. 45.

Consequently each square will be the area unit.

On this squared paper we will draw (Fig. 45) a square whose side will contain 7 divisions, so that the length of this side has 7 for its measure. The squares contained in this figure are made up of 7 rows, each containing 7 squares; their total number then is 7 times 7, or $7 \times 7 = 7^2 = 49$. And using a to indicate the number of divisions on the side of a square (whether this number be 7 or any other number), the area of the square would be $a \times a = a^2$; that is to

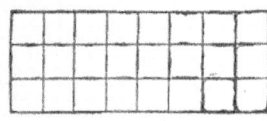

FIG. 46.

say, the number which measures the area of the square is the 2nd power of the number which the side measures. It is for this reason that the square of a number is called its 2nd power.

We will take now (Fig. 46) a rectangle whose sides are 8 and 3; the number of squares, that is to say, the number which will measure its area, will be 8×3; if, instead of 8 and 3, we have a and b, the area of the rectangle will be measured by the product ab.

Now imagine (Fig. 47) a parallelogram ABCD, and take its height CH.   If, having formed this parallelogram in cardboard, for instance, we cut off, by a line along CH, the triangle CHB which is shaded on the figure, and if we set down this triangle on the left, laying CB on DA, we form the rectangle CDKH, whose area will be the same as that of the parallelogram, since it is made up of the same pieces.   This rectangle has for its sides the base CD of the parallelogram, and its height CH.   Then the area of a parallelogram has for its measure the product of the numbers which measure its base and its height.

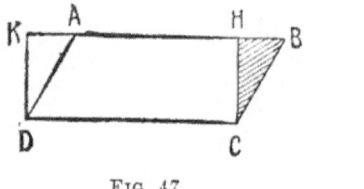 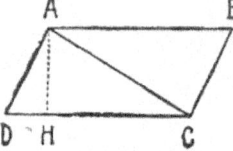

FIG. 47.   FIG. 48.

A parallelogram (Fig. 48) being cut in two by a line along the diagonal AC, the two triangles CBA, ADC will lie exactly one upon the other.   Then the parallelogram has an area double that of the triangle ADC, and the latter has an area half of that of the parallelogram.   Making the product of the base DC, by the height AH, and taking the half of this product, we shall have the number measuring the area of the triangle.

A trapezium (Fig. 49) can be split up in the same way into two triangles.   We deduce from this that to obtain the number which measures its area, it is necessary to multiply the height AH by the half of the sum of its bases AB, DC.

We can also (Fig. 50, A) transform a trapezium into a rectangle of the same area.   It is only necessary to draw the heights HK, IJ through the centres L, M, of the sides AD, BC.   The triangular shaded portions LDH, MCI, can be laid exactly on LAK and MBJ, and thus form the rectangle. This shows us that HJ, or LM, is equal to half of the sum of the bases AB, CD.

Finally (Fig. 50, B), if we prolong the side AB by a length BF, equal to DC, and the side DC by a length CG, equal to AB, the figure AFGD is a parallelogram; and cutting it along BC we have two

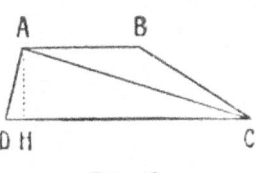

Fig. 49.

trapeziums, which will fit over one another; each of them is then the half of the parallelogram, which gives

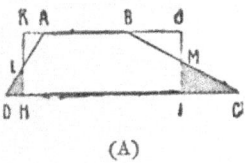

(A)

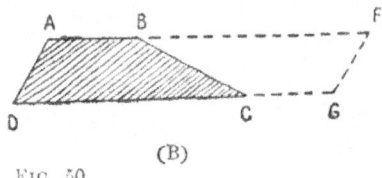

(B)

Fig. 50.

us yet again the area of a trapezium by a new means, merely by a simple cut of the scissors across the cardboard parallelogram.

We can summarise what has gone before in the following formulæ :—

Square	Side A	Area A^2
Rectangle	Sides A, B	Area AB
Parallelogram	Base A ; height H	Area AH
Triangle	Base A, height H	Area $\dfrac{AH}{2}$
Trapezium	Bases A, B, height H	Area $\dfrac{(A + B) H}{2}$

Moreover, as soon as the pupil can measure the area of a triangle, he can determine that of any polygon whatever, ABCDEF (Fig. 51), since by the diagonals AC, AD, AE, starting from a corner, he can cut the polygon into the triangles ABC, ACD, ADE, and AEF.

To determine in this manner exact areas, such as those of a door, a window, a table, the floor or the ceiling of a room, a playground, etc., much practice will be necessary. A tape measure will be sufficient to use for this purpose.

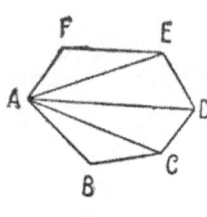

FIG. 51.

According to the object, he will take for the unit of length the yard, the foot, the inch, the metre, the decimetre and the centimetre ; without having up to this point any idea of theory, in this manner he will familiarise himself with the most simple applications of our weights and measures and of the metric system ; he will grasp them intuitively ; he will have an exact idea of the employment of the various units ; and this acquisition, already useful in itself, will become later a very considerable help when he really begins his studies.

24. The Asses' Bridge.

There is in geometry a proposition at once celebrated and important, but which has been the despair of many generations of scholars, because the academical demonstration that is usually given of it is hardly natural, and difficult to remember. It is known under different names ; it is called " the square of the hypotenuse," " the theorem of Pythagoras " (although it was known many centuries before Pythagoras), lastly " the asses' bridge," undoubtedly because ordinary scholars stumble at it and have some trouble in getting over it.

We already know what a rectangular triangle is. The greatest side BC (Fig. 52), that which is opposed to the right angle, is called the *hypotenuse*. If three squares be formed, BDEC, CFGA, AHIB, having for their sides the hypotenuse and the two other sides, the area of the first will be equal to the sum of the areas of the two others. This is the opening statement of the famous asses' bridge.

Now, there is a very simple method of verifying this proposition, a method which was invented in India in the very earliest times and can be used for a most exact demonstration when the study of geometry has been begun —though as yet we have not entered on it.

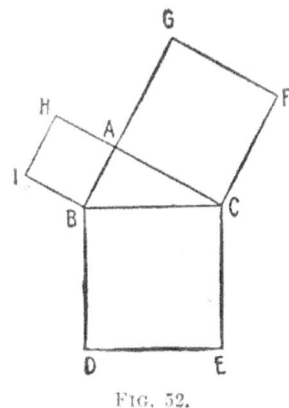

FIG. 52.

Let us take a square (Fig. 53 (1)) whose side is AB. Marking a point C between A and B, we will construct, either in wood or cardboard, rectangular triangles having AC and CB for their sides which contain the right angle. Four will be sufficient. Arrange them, numbering them 1, 2, 3, 4, as they are shown in Fig. 53 (1), where the shaded parts represent these little triangles. We see that they form a pattern which allows a square to be seen in the interior, which has for its side exactly the hypotenuse. This square is then what remains when a part of the large

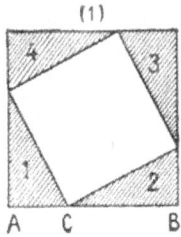

(1)

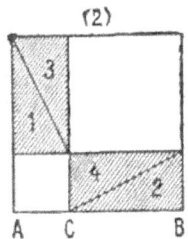

(2)

FIG. 53.

square has been covered with the four triangles. Now, let us slip our four triangles into the position indicated by Fig. 53 (2). What now remains is two squares, the two squares constructed on the sides of the right angle.

M   F

Then both of them have the same area as the square of the hypotenuse in Fig. 53 (1). It is just a very simple game of patience; the child who has practised it once or twice will never forget it in his whole life, and will never be dismayed nor confused when he approaches the asses' bridge. The greatest blunder of all is to complicate simple things and to make difficult anything which is easy.

25. Various Puzzles; Mathematical Medley.

On a segment ABC (Fig. 54) let us construct a square ACIG; then taking CF = BC, let us draw FED parallel to AC; also draw BEH parallel to CI. The large square will be cut into four parts by the lines BH and FD; this can be done by two snips of the scissors. These four pieces are :

    1st BCFE, square having for its side BC.

    2nd EHGD, square having for its side DE which is equal to AB.

    3rd EFIH, rectangle having its sides equal to AB, BC.

    4th ABED, rectangle like the preceding one.

We have just verified this theorem of geometry :

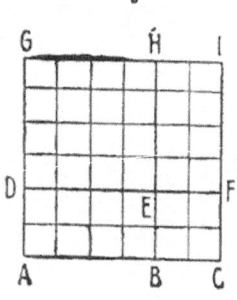

Fig. 54.

"The square constructed upon the sum of two lines is equal to the square constructed upon the first, plus the square constructed upon the second, plus twice the rectangle constructed with these two lines as sides."

If we have drawn the figure on squared paper, by estimating the areas of all these figures, that is to say, counting the divisions, we have the proposition of arithmetic :

"The square of the sum of two numbers is equal to

the sum of the squares of these two numbers, plus twice their product.''

If we indicate AB by a, BC by b, we have finally this formula in Algebra :

$$(a + b)^2 = a^2 + 2ab + b^2.$$

These are three truths which are loaded on the memory of the unprepared child three times, while in reality they are all one thing which jumps to the eye. Its appearance, its dress, is different, but it is itself the same in each case. Knowing this beforehand, he will be spared loss of time and vain efforts, and, more than all, will know that these classifications are

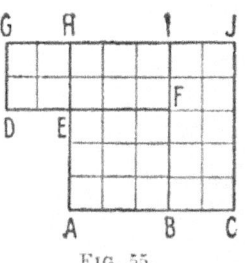

FIG. 55.

necessary, but often artificial by the force of things; and will accustom himself early to recognise the analogies he will encounter.

We are going to mention others. Let us form (Fig. 55) on the segment ABC a square having for one of its sides

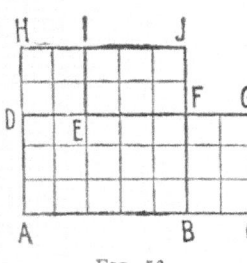

FIG. 56.

AB, which square will be ABFE, and a square ACJH, having a side AC. Produce BF to I, and on EH construct the square EHGD.

For the square ABFE, we must take away from the whole figure the rectangles BCH, FIGD; the whole figure is made by the reunion of two squares whose sides are equal to AC and BC; the two rectangles are similar, and their sides are equal to AC and BC; finally AB is the difference of AC and BC. Then :—

Geometry.—The square constructed on the difference of two segments is equal to the sum of the squares con-

F 2

structed on these two segments, less twice the rectangle
constructed with the two segments as sides.

Arithmetic.—The square of the difference of two numbers
is equal to the sum of the squares of these numbers, less
twice their product.

Algebra.—This formula will be $(a - b)^2 = a^2 - 2ab + b^2$.

One more example (Fig. 56); ABJH is a square, ACGD
a rectangle ; FG, FJ, DE are equal to BC, DEIH is a
square.

The rectangle ACGD has thus for sides AB + BC and
AB − BC ; as the two rectangles BCGF, FJIE are
identical, by taking away the first and putting it in the
place of the second we shall have ABJIED, which is
the difference of the squares ABJH, DEIH constructed
on AB and DE = BC.   Therefore :—

Geometry.—The rectangle having as sides the sum and
the difference of two segments is equal to the difference
of the squares having these two segments as sides.

Arithmetic.—The product of the sum of two numbers
by their difference is equal to the difference of their
squares.

Algebra.—This formula will be $(a + b)(a - b) = a^2 - b^2$.

And to verify so many propositions, concerning so
many sciences, it will only be necessary to cut some
shapes of cardboard into pieces, after having made the
figures with great care.

These games of cutting up cardboard have sometimes
been called brain puzzles.   This is very unjust, because
used in the manner we have just indicated they prevent
on the contrary much puzzling of the brain in the future
by dint of teaching by means of the eye.

26. The Cube in Eight Pieces.

Let us take (Fig. 57) a wooden cube, and starting
from one of the corners O, let us lay out, on the three
edges which end there, three lengths equal to each other,

OA, OB, OC. Suppose that we saw through along AAA, BBB, CCC, at each of the three points thus obtained.

By this means the cube is cut into eight pieces. To make the explanation easier, by means of looking at the object itself, let us call (Fig. 57) the length DA a, and the length AO b, constructing thus Fig. 58. The two parts of which it is composed represent what we see after the cuts along AAA, and BBB, when we look at the cube from above. As well as this, the letters (a) (b) between brackets show the thickness after the cut along CCC. The left figure shows what is underneath CCC, and that on the right what is above.

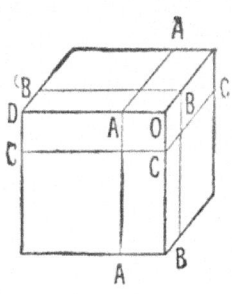

FIG. 57.

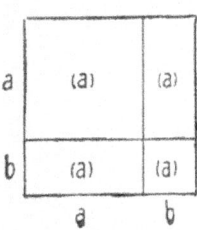
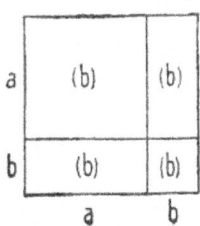

FIG. 58.

We can easily see that we shall have eight parallelopipeds whose dimensions will be :

Figure on the left, *aaa, aba, bba, baa.*
Figure on the right, *aab, abb, bbb, bab.*

That gives us then :

a cube whose edge is a ;
" " " b ;
3 parallelopipeds having for sides a, a, b ;
3 " " " a, b, b ;

The edge of the cube which we have cut into eight pieces was $a + b$.

We verify thus that the cube constructed upon the sum of two segments a, b is made up :—

First, of the sum of the cubes constructed upon each of the segments ;

Second, of three times a parallelopiped, having for its base a square with the side a, and for its height b ;

Third, of three times a parallelopiped, having for its base a square with a side b, and for its height a.

This is Geometry.

The same figure shows us that in Arithmetic the cube of the sum of two numbers is equal to the sum of the cubes of these two numbers, plus three times the product of the first by the square of the second, plus three times the product of the second by the square of the first.

Finally (Algebra) this gives the formula :

$$(a + b)^3 = a^3 + 3a^2b + 3ab^2 + b^3.$$

This is quite analogous to what we have done for the square of a sum in the preceding section.

With a sufficient number of little wooden cubes, the constructions which we have indicated may be made, and also many more. These are games which, directed with a little method, help the child very much to see the figures in space, and engage his attention.

If necessary, the cutting up of the cube might be done by means of a piece of soap taken from a bar, cutting it carefully with a wire instead of using a saw. But the wooden cube is much to be preferred, and is certainly neither difficult nor expensive to procure.

27. Triangular Numbers.—The Flight of the Cranes.

Edward Lucas attributes the origin of the numbers which have been called *triangular* to the observation of the flight of certain birds. At the head there flies a single bird ; behind him, on a second line, there are two ; on a third line behind them, there are three, and so on ; so that the general disposition of the flying column presents the appearance of a triangle.

It is easy to give ourselves a precise idea of these numbers and to represent them on a squared pattern; looking at Fig. 59, for example, and considering, to begin with, the part A only, which shows us, on the top, one division, then two divisions in a second row, then 3, 4, 5, 6, 7 divisions in the following rows, up to the seventh.

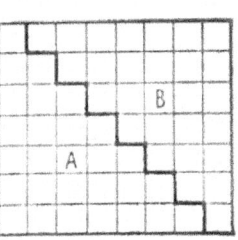

FIG. 59.

We have then the 7th triangular number

$$1 + 2 + 3 + 4 + 5 + 6 + 7 ;$$

to find its value, we can add up, which will give us 28. But that would teach us nothing about any other triangular number. If we wanted to have the 1,000th, for instance, we would have to add from 1 to 1,000, which would be long and very wearisome. Instead of that, let us now look at the whole of Fig. 59 ; the part B, if we look at it from the bottom to the top, or if we turn it upside down, shows us still, by the number of its squares, the same triangular number. The entire figure then represents twice the triangular number in question ; and as it is composed of seven rows, each having eight divisions, the total number of divisions is 7 × 8, and the required number will be the half of this product, that is to say, 28.

We shall have, in other terms,

$$1 + 2 + 3 + 4 + 5 + 6 + 7 = \frac{7.8}{2} = 28.$$

If we wanted to get the 1,000th triangular number, supposing that we did it the same way, we should have

$$1 + 2 + 3 + \ldots + 1{,}000 = \frac{1{,}000 \cdot 1{,}001}{2} = 500{,}500.$$

This is shorter than adding.

And as, instead of 1,000, we might have any whole number n, we have also

$$1 + 2 + 3 + \ldots + n = \frac{n(n+1)}{2},$$

which expression will allow us to find the nth triangular number, which we can call T_n.

The total number of divisions in Fig. 59 is 2 T_n. If we take away the last column, a square of seven lines will remain, each containing 7 divisions. We see, therefore, that the new figure is formed of the combination of the triangular numbers T_6 and T_7. We have therefore

$$2\,T_7 - 7 = 7^2 = T_7 + T_6.$$

If we add below a new row of eight divisions, we see that we have

$$2\,T_7 + 8 = 8^2 = T_8 + T_7,$$

just by looking at the figure.

And as, in place of 7, we might have taken any other number n

$$2\,T_n - n = n^2 = T_n + T_{n-1}$$

$$2\,T_n + n + 1 = (n+1)^2 = T_{n+1} + T_n.$$

These formulæ, which appear very learned, do not, however, require even the least calculation, since the pupil can *read* them from figures, since he can *see* them, since he can make them with his hands by means of little wooden squares, or even with simple counters by placing one in each division.

28. Square Numbers.

Take (Fig. 60) a square, composed of 7 rows, of 7 divisions each, in all $7 \cdot 7 = 7^2 = 49$ divisions. On this figure, by means of traced lines, we see the successive squares, 1, 2^2, 3^2, 4^2, 5^2, 6^2 divisions.

The first square, of 1 division, is represented by the division at the top left-hand side. To pass from this square to that of 2^2 or 4 divisions, we notice that it is necessary to add 3 divisions, so that $1 + 3 = 2^2$; to pass to the foliowing square, of 9 divisions, we must add two from the right, two underneath, and one from the right below, which makes 5; and by continuing in the same manner, we see that

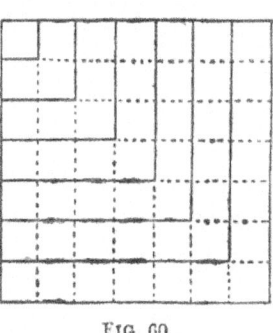

FIG. 60.

$$7^2 = 1 + 3 + 5 + 7 + 9 + 11 + 13.$$

Which means that the square of 7 is equal to the sum of the first 7 uneven numbers.

Instead of 7, let us take any whole number we like, n. The first uneven numbers are 1, 3, 5 . . . and the nth

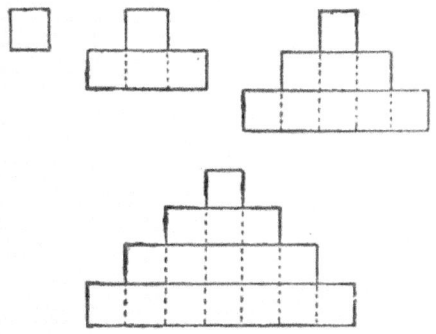

FIG. 61.

is $2n - 1$. As we have been able to make this figure up to the number n, we have

$$n^2 = 1 + 3 + 5 + \ldots + 2n - 1,$$

and this only expresses what we see in Fig. 60.

We thus see that there is another way by which we can represent square numbers ; it is shown in Fig. 61, where we see the squares of 1, of 2, of 3, and 4. With little wooden squares it will be easy to make and also change these various figures.

Without any trouble, we are now going to solve a very much more difficult problem, that of finding the sum of the squares of 1, 2, 3, 4, for instance. By making use of Fig. 60, and by laying the squares of 1, 2, 3, 4 from bottom to top, we have at once Fig. 62, which needs

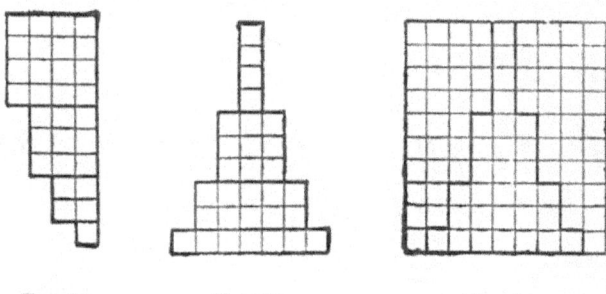

FIG. 62.   FIG. 63.   FIG. 64.

no explanation. Making use of the various elements of Fig. 61, we see that there are

    4 rows of 1 division
    3 ,,   3 divisions
    2 ,,   5   ,,
    1 ,,   7   ,,

which, placed each under the other, give Fig. 63.

Let us bring together Fig. 64, Fig. 62, the same turned over, and also Fig. 63. We obtain a rectangle which will contain three times the required number of divisions.

The number of rows in this rectangle is

$$1 + 2 + 3 + 4, \text{ or } \frac{4.5}{2} = 10.$$

The number of divisions contained in each row is, as we can see on the first line,

$$4 + 1 + 4, \text{ or } 2.4 + 1 = 9.$$

The total number of divisions then is $10.9 = 90$, and the required number will be the third of 90, that is to say, 30. We thus verify that

$$1^2 + 2^2 + 3^2 + 4^2 = 1 + 4 + 9 + 16 = 30.$$

But if, instead of 4, we had taken any number whatever, n, and if we had done exactly the same thing, the rectangle of Fig. 64 would have

$$1 + 2 + 3 + \ldots + n, \text{ or } \frac{n(n+1)}{2} \text{ lines,}$$

and $2n + 1$ columns.

The total number of its divisions would be then

$$\frac{n(n+1)(2n+1)}{2},$$

and to have the required number we must take the third of it, which shows us that

$$1^2 + 2^2 + 3^2 + \ldots + n^2 = \frac{n(n+1)(2n+1)}{6}.$$

This is a formula which candidates at the Polytechnic School cannot always prove, giving themselves an infinite amount of trouble and endless calculation, whilst we establish it by amusing ourselves with little wooden squares like those we have already seen.

This determination of the sum of the squares of the first n numbers had formerly an important practical application in artillery, when spherical projectiles were in use (cannon-balls or shells). Indeed, these were often arranged in arsenals, by forming a square on the ground, then, above, another smaller square, and so on up to the top, which was made of a single ball or shell. This was called a *pile of balls with a square base.* Then, in order to count the balls contained in a pile, it suffices to count the number n of the balls on one side of the base and to apply the above formula. For example, if n equals 17, the required sum is

$$\frac{17.18.35}{6} \text{ or } 1785.$$

The child could amuse himself by forming piles of oranges in the same way, provided that they are about the same size, or, more simply perhaps, by using billiard balls, laying them on a light bed of sand to keep them from rolling and so spoiling the erection.

29. The Sum of Cubes.

To represent a number raised to the cube, such as $2^3 = 2.2.2$, $3^3 = 3.3.3$, etc., it would be convenient to have a large number of little wooden cubes, rather larger than dice, which would serve both to make the figures of which we have spoken above, and also to perform the various operations which are to follow.

We need not be quite so strict, however, for we can dispense with the cubes, and replace each unit by a small

FIG. 65.

flat square, of wood or cardboard, or even by a simple counter. It is this last supposition that we shall adopt. Later, when we have seen how easy the constructions are, they will be made all the more readily, by means of squares or cubes; he who can do the most can certainly do the least.

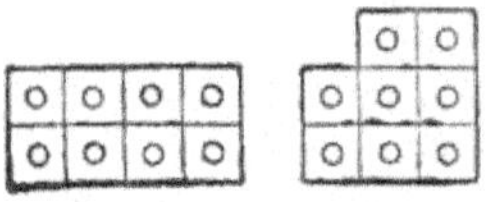

Begin by seeing how, with our counters, we can represent successive cubes. The cube of 1 is 1; a single counter will represent it.

The cube of 2 is $2 \times 2 \times 2$, or 8; therefore it will be composed (Fig. 65) of 2 squares of 4 counters each, the squares placed side by side, in the first part of the figure. But, as in the second part, these 8 counters may be arranged in another manner, by keeping the first three columns, and by placing the fourth, which has become horizontal, on top.

Let us proceed to the cube of 3, which is $3 \times 3 \times 3$,

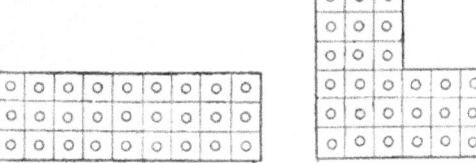

Fig. 66.

or 27 ; it is represented (Fig. 66) by 3 squares of 9 counters each, side by side, in the first part of the figure.

The second part is obtained by keeping the first six

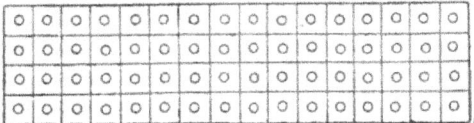

Fig. 67.

columns and placing on top the last three, which are now horizontal.

Finally, for the cube of 4, we will do the same, by keeping

(Fig. 67) the first 10 columns of the first part and placing above them the last six, horizontally, to obtain the second part of the figure.

If we now put together (Fig. 68) the second parts of Figs. 65, 66, 67, by adding a counter to the left at the top, which will represent the cube of 1, we shall have the sum of the cubes of 1, 2, 3, 4 in the form of a square, in which the number of the counters in a row or a column is $1 + 2 + 3 + 4$, or 10. The sum of these cubes is then 100.

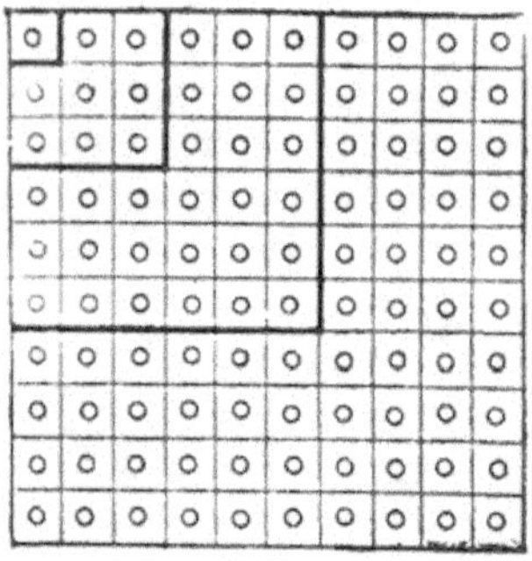

FIG. 68.

It will be interesting to take counters of different colours to represent each of the cubes. The figure will then be so much the more striking.

The method of construction indicated might thus be carried out as far as the cube of any number whatever, n, and shows that *the sum of the cubes of the n first whole numbers is equal to the square of the sum of these numbers.*

This is expressed by the formula

$$1^3 + 2^3 + 3^3 + \ldots + n^3 = (1 + 2 + 3 + \ldots + n)^2.$$

It is an expression which can also be written $(T_n)^2$ or

$$\left(\frac{n(n+1)}{2}\right)^2.$$

By this means we have a result which is much more troublesome and difficult to obtain by calculation. Here we arrive at it by a simple construction game[1]!

---

[1] It is perhaps worth noticing that in our table of multiplication without figures (Fig. 12) we find exactly that the successive cubes 2^3, 3^3, . . . represent the number of divisions in the spaces separating the squares of $1, 1 + 2, 1 + 2 + 3. . . .$ Indeed, strictly, we should have been able, with this simple table, to demonstrate all that we have just seen here.

30. The Powers of 11.

If we take the number 11, and wish to form the square of it, the multiplication will be very easy

$$
\begin{array}{r}
11 \\
11 \\
\hline
11 \\
11 \\
\hline
121
\end{array}
\qquad 11^2 = 121.
$$

To obtain the cube we shall have

$$
\begin{array}{r}
121 \\
11 \\
\hline
121 \\
121 \\
\hline
1331
\end{array}
\qquad 11^3 = 1331.
$$

The fourth power would need the multiplication as below

$$
\begin{array}{r}
1331 \\
11 \\
\hline
1331 \\
1331 \\
\hline
14641
\end{array}
\qquad 11^4 = 14641
$$

Let us fix our attention upon these figures

1, 2, 1 ;   1, 3, 3, 1 ;   1, 4, 6, 4, 1,

which are employed to write the powers.

We would have been able to have them, with less writing, without putting down the multiplications, if we notice first that we begin and end with 1 ; and also that we have only to add two figures which follow each other to get a figure of the following power.

Thus from 11 we get 121, because $1 + 1 = 2$;

From 121 we get 1331, because $1 + 2 = 3$, $2 + 1 = 3$;

From 1331 we get 14641, because $1 + 3 = 4$, $3 + 3 = 6$, $3 + 1 = 4$.

These remarks have led to a means of obtaining these figures (and many other numbers) very easily, as we are going to see in the following section.

It is all the more useful because the numbers of which it treats are of great importance in algebra, where the pupil will have to deal with them later, however little he may study mathematics.

31. The Arithmetical Triangle and Square.

Let us write (Fig. 69) the figure 1 as many times as we like each under the other. Suppose that at the right of the first one, on top, there are noughts, which it is not necessary to write. We form the second line adding 1 and 0, which makes 1, and writing this 1 to the right of that which is already put down. Let us pass on to the third line; we read in the second, 1 and 1 making 2, which we write; then 1 and 0, 1, which we place to the right of the 2; similarly, starting from the 3rd line we form the 4th, 1 and 2, 3; 2 and 1, 3; 1 and 0, 1. And so on as far as we like. The first rows of the figure give us the numbers that we have already found to be the powers of 11.

1							
1	1						
1	2	1					
1	3	3	1				
1	4	6	4	1			
1	5	10	10	5	1		
1	6	15	20	15	6	1	
1	7	21	35	35	21	7	1

FIG. 69.

This figure is called, from the name of its illustrious inventor, *the arithmetical triangle of Pascal.*[1]

[1] Blaise Pascal, French scholar and man of letters, born at Clermont-Ferrand (1623—1662).

The same numbers appear in a figure which is in no way different to the preceding one except by its arrangement (Fig. 70).   The figure 1 is written in each of the divisions of a first row and in those of the first column of a piece of squared paper.   Afterwards all the other divisions are filled successively by putting in each the sum of the number which can be read above and the one which can be read at the left.

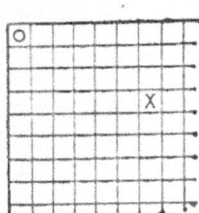

1	1	1	1	1	1	1	1
1	2	3	4	5	6	7	8
1	3	6	10	15	21	28	36
1	4	10	20	35	56	84	
1	5	15	35	70			
1	6	21	56				
1	7	28	84				
1	8	36					

FIG. 70.

Here the figures 1, 1 ; 1, 2, 1 ; 1, 3, 3, 1 ; . . . appear, no longer in the horizontal lines, but in those which are oblique ascending from left to right.

This (Fig. 70) is called *the arithmetical square of Fermat.*[1]

If we consider (Fig. 71) an ordinary chess-board, the left corner division O and any division whatever X, we can ask ourselves by how many different ways it is possible to go from O to X without going back, that is to say, moving always from left to right and downwards. The arithmetical square of Fermat gives us the answer if we lay it on the chess-board.   Thus for Fig. 71 as it is drawn we have 84 different ways of reaching X from O.

FIG. 71.   The numbers of these figures possess many very curious properties.   But it would be premature to consider them now.

---

[1] Pierre Fermat. French mathematician, born at Beaumont de Lomagne (1601—1665).   He possessed probably the most powerful genius from an arithmetical point of view that has ever been known.

32. Different Ways of Counting.

When we began (Section 3) to make numbers by means of little sticks, then bundles, faggots, etc., which leads to numeration, we could equally well take any other number than 10 little sticks to make a bundle.

For example, we might have arranged that 8 little sticks would make a bundle, 8 bundles a faggot, and so on. The result would have been that the figures necessary to write any number (Section 10) would have only been 1, 2, 3, 4, 5, 6, 7, to which, of course, the nought would have to be added.

Such a method of writing numbers is what is called a *system of numeration*, and the number chosen is called the *base* of this system.

Thus the system of which we have so far made use, which is universal, is called the decimal system, and has for its base 10. The one which we have just indicated would have 8 for its base, and might be called the octesimal system.

Supposing that 12 were taken as the base of a system, it would be called the duodecimal system, and it would take 12 little sticks to make a bundle, 12 bundles to make a faggot, and so on. There would have to be then, excluding the nought, eleven figures, that is to say, the 9 of the decimal numeration, and two others to represent 10 and 11.

When a numeration system has for its base a number B, it always requires B—1 figures, without counting the nought, and the number B is invariably written as 10.

It is useful to know how to put down a figure in one system of numeration when it is given to you written in another, and really it is perfectly easy.

For instance, take 374, written in the system with a base 8. We will try to write it in the decimal system.

If we think of our little sticks we can see that the number in question contains

4 little sticks	..	..	..	..	4
7 bundles of 8 little sticks		..	..	56
3 faggots of 8 × 8 little sticks..			..	192

$$252$$

In practice we would arrive at the same result more quickly by starting from the left and saying : 3 faggots of 8 bundles, plus 7 bundles, gives 31 bundles ; 31 bundles of 8 little sticks makes 248, which, with 4 added, makes 252.

If, on the contrary, the number 598 is written in the decimal system, and we wish to have it expressed in the system with 8 for base, there will be nothing easier than to take away 8 as many times as we can, and the remainder will be the last figure to the right. So we divide 598 by 8, and take the remainder, 6; this operation gives us the number of bundles of 8, which is 74; dividing by 8 we have the number of faggots, 9, and for remainder, 2 bundles ; 2 is the 2nd figure. Dividing 9 by 8 we see finally that we have for remainder 1 bundle (1 is the 3rd figure) and that we have 1 box (1 is the 4th figure).

This will be expressed thus :

```
598 | 8
 38   74 | 8
  6    2   9 | 8
           1   1
```

and 1126 is the required number, written in the system of base 8.

If it was necessary to write this number with a 12 base, we should have

```
598 | 12
118   49 | 12
 10    1    4
```

and the result would be 41(10), representing by (10) the figure 10 of the duodecimal system.

We have just seen that 374 (system 8) is expressed as 252 (decimal system).   In the system base 12 it would be written 190.

We go from one system to another at will, using the decimal system as an intermediary.

With use, the pupil can calculate in any system whatever, only the essential point is this, never to let him forget that carrying is no longer done by tens, but by groups of B, if B is the base ; this naturally calls for some practice.

We give below the number 1000 in the decimal numeration, written in the numeration systems with various bases 3, 4, 5, . . . up to 12.

B = 3	..	..	1101001
4	..	..	33220
5	..	..	13000
6	..	..	4344
7	..	..	2626
8	..	..	1750
9	..	..	1331
10	..	..	1000
11	..	..	82(10)
12	..	..	6(11) 4

It is worthy of notice that, using the numeration system with base 3, and employing negative figures, the numbers reduce themselves then to 0, + 1, − 1.   This fact is rendered more interesting when we learn that it can be put to practical use in certain questions relative to hydraulic lifts.[1]

M. Marcel Deprez (membre de l'Institut), to whom we owe the transport of energy by electricity, has been good enough to tell me of a curious way of weighing by means of a balance.   Let us suppose that we place weights in both scales.   Given these conditions, the problem pro-

[1] This application, in a previous French edition, was placed at the end of the book, under the title *Note on a question of weighing.*

posed is to determine a system of weights (a single weight of each kind) starting from 1 gramme, let us say, in such a manner that it will be possible to balance bodies weighing 1, 2, 3 . . . grammes up to a determined limit.

We see that with the 2 weights 1 gramme and 3 grammes we can weigh up to 4 grammes, since $2 = 3 - 1$, and $4 = 3 + 1$.

By taking the 3 weights, 1, 3, 9 grammes, we can weigh up to 13 grammes.

In general, if we have taken the n weights 1, 3 . . . , 3^{n-1} grammes, we can weigh up to $\dfrac{3^n - 1}{2}$ grammes. For instance, with the 7 weights, 1, 3, 9, 27, 81, 243, 729 grammes, we can weigh from 1 up to 1093 grammes.

This question, as we might note, leads back to the writing of successive numbers in the system of base 3 by utilising negative figures. Then, instead of the figures 1, 2, we use 1, $\bar{1}$; and $\bar{1}$ shows that the corresponding weight ought to be placed in the second pan of the balance. [$\bar{1}$ is another way of writing $- 1$.]

For instance, 59 is written in this system $1\,\bar{1}\,1\,1\,\bar{1}$, for $59 = 81 - 27 + 9 - 3 - 1$. To weigh 59 grammes we will put the weights 81 and 9 in a scale, and 27, 3, 1 in the opposite scale ; adding to this last a body weighing 59 grammes, the balance will be in equilibrium.

It may be interesting to add here some observations on Roman numeration. It is now no longer used except for the purpose of marking the hours on the dials of watches or clocks. The child will be able also to decipher the dates on old inscriptions if he understands it, but that is all, so that the actual mathematical interest is of a very moderate kind. It is quite different from the teaching point of view. I must content myself with only a summary of the observations which M. Godard (the then director of the school, 'Ecole Monge) brought to my notice many years ago.

If sticks are laid in order on a black table, for instance, taking care to space them equally, and we suddenly ask anyone, without any warning, how many sticks there are in a certain group, the answer will be given immediately if the group contains two, three, or four; beyond that, for five and upwards, there would have to be a preliminary rapid operation of the mind, a mental decomposition of the number, so that the answer would no longer be the result of visual impression. This is a well-assured fact, which is verified by many experiments.

On the other hand, the number five plays an important part in Roman numeration.

We are inclined to ask if it has not originated from the physiological fact that we have just pointed out, and its symbols of expression from the anatomical disposition of the human hand.

The numbers one, two, three, four would be represented by one, two, three, four fingers raised :

I, II, III, IIII.

We can make a tolerably good imitation of the shape of the letter V by means of the whole hand held up, the thumb stretched away from the four fingers.

Ten can be shown by the joining of two hands, one upwards, V, the other downwards, Λ, which gives the letter X.

Only the first principles of Roman numeration have been mentioned, and we refrain from any comment on the other symbols L, C, M. . . . It is, however, useful to notice that the consecutive repetition of the same sign more than four times is to be avoided.

To obtain the numbers between five and ten the necessary units follow the sign V :

VI, VII, VIII.

Similarly, for numbers above ten, we have

XI, XII, XIII.

It is probable that as a later, but still very early improvement, the idea was carried out of showing subtraction by placing the unit (or other symbol) *to the left* of a fixed sign, while transferred *to the right* the same figure would signify addition. Thus the expressions

IV, IX, XL, . . .

give us five less one, or four; ten less one, or nine; fifty less ten, or forty; there are others of a similar kind. It is remarkable to notice here, in however embryonic a form, this first attempt at graphic translation of the *sign* by the *sense*.

These observations seemed to me sufficiently curious to deserve mention. They seem to result in the idea that Roman numeration was one with a base 5, but incomplete, since in it different symbols were not used to indicate the first four numbers, and, above all, because that central pivot of all rational numeration—that *nothing* which is *everything* in arithmetic—that invaluable resource, the nought, is lacking.

33. Binary Numeration.

We have seen, in the last section, that if B is the base of a system of numeration, this system requires $B - 1$ figures, without the nought. If 2 is the base, only one figure will be required, the figure 1.

The idea of this numeration, in which all numbers are written by means of only two characters, 1 and 0, seems to belong to Leibnitz,[1] although it is said that the Chinese made use of it in ancient times.

For ordinary use in calculations this system would be inconvenient because of the length of its expressions. Thus the number 1,000 in the decimal numeration would be represented in the binary system by the figures 1111101000, ten in all. But in certain scientific applications the employment of the binary system is at once useful and interesting. As well as this, we find in it the explanation of various games, such as " the ring

[1] Leibnitz, German philosopher and mathematician, born at Leipzig (1646—1716).

puzzle " and Hanoi Tower, and also a little drawing-room
game which depends upon the use of binary numeration,
and of which Edward Lucas gives us a description in
his " *Arithmetique amusante* " under the name of " *The
Mysterious Fan.*"

To make this clear, suppose that we have written the
31 first numbers in binary enumeration :

1	1				
2	10	12	1100	22	10110
3	11	13	1101	23	10111
4	100	14	1110	24	11000
5	101	15	1111	25	11001
6	110	16	10000	26	11010
7	111	17	10001	27	11011
8	1000	18	10010	28	11100
9	1001	19	10011	29	11101
10	1010	20	10100	30	11110
11	1011	21	10101	31	11111

Now, on a piece of cardboard, A, let us write (decimal
system) all of these numbers which end in 1 (binary
system) :

A	B	C	D	E
1	2	4	8	16
3	3	5	9	17
5	6	6	10	18
7	7	7	11	19
9	10	12	12	20
11	11	13	13	21
13	14	14	14	22
15	15	15	15	23
17	18	20	24	24
19	19	21	25	25
21	22	22	26	26
23	23	23	27	27
25	26	28	28	28
27	27	29	29	29
29	30	30	30	30
31	31	31	31	31

On a second piece of cardboard, B, we write down also the numbers whose 2nd figure, starting from the right, is a 1 in binary numeration; then the same (cardboard slips C, D, E) for the 3rd, 4th, and 5th figures.

If you give these five slips to anyone, and ask him to think of a number thereon, and to hand you the slips on which his number is written, but only those—that will tell you the numbers which enable it to be written in binary numeration. It is quite easy to verify that, in order to find the number selected, we have only to add the first numbers written on each slip. Suppose we take 25 as the number chosen; the slips you will receive will be A, D, E, beginning with the figures 1, 8, 16; $1 + 8 + 16 = 25$.

The game may be played with the number 63 instead of 31, using 6 slips instead of 5, and with 127 also (this latter requiring 7 slips). The apparent divination may be rendered still more mysterious by the use of proper names. It is only necessary to make a list, giving a number to each name, and writing the names on the slips, remembering that the various slips begin with 1, 2, 4, 8, 16, 32, 64.

Anyone can make this group of slips of cardboard, as far as 7 for instance, and even if the performer does not gain a reputation as a sorcerer, at least there will be the solid satisfaction of having plenty of practice in rapid and accurate mental addition. Naturally, without both quickness and accuracy, the aforesaid reputation can never be maintained.

34. Arithmetical Progressions.

Take a series of numbers,

$$4 \quad 7 \quad 10 \quad 13$$

for example, such that the difference between two consecutive ones is always the same:

$$7 - 4 = 10 - 7 = 13 - 10 = 3.$$

Such a series is called *a progression by difference* or *an arithmetical progression*.

The difference, being constant, 3 in this example, is called the *common difference* of the progression.

The numbers 4, 7, 10, 13 are the *terms* of the progression. Here we have only written four, but we could have as many as we liked.

It may be pointed out that the series of integral numbers 1, 2, 3 . . . forms an arithmetical progression with common difference 1, and the series of the odd numbers 1, 3, 5 . . . forms an arithmetical progression with the common difference 2.

We will try and represent graphically (Fig. 72) the progression 4, 7, 10, 13, the example mentioned above. On squared paper, counting 4 divisions on the first line

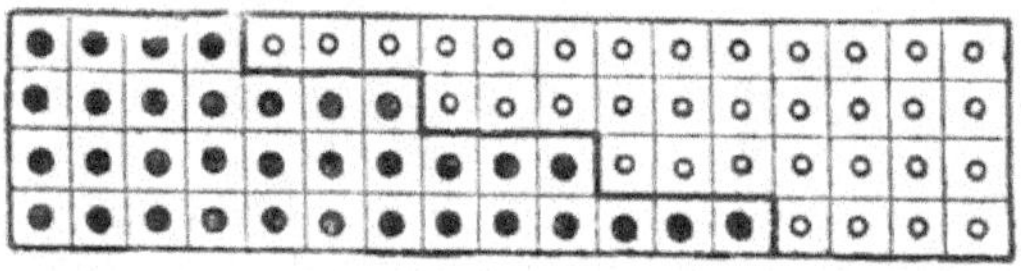

Fig. 72.

we place black counters in each; taking 7 divisions on the 2nd line, 10 on the 3rd, 13 on the 4th, we fill each division with a black counter.

Then, by adding 4 divisions to this last line, and completing the rectangle, we see that this rectangle contains the terms of the progression twice over, as shown by the black and the white counters.

We verify in this way that the sum of extreme terms is the same as that of two terms equidistant from the extremes. Finally, the sum of the terms of the progression will be half the number of the divisions of the rectangle, that is to say $\dfrac{17 \times 4}{2}$.

As a rule, if a and b are the two extreme terms and if

n is the number of the terms, this sum is expressed by $\dfrac{(a + b)n}{2}$.

If $a = 1$, and if the common difference is 1, then $b = n$, and we have $\dfrac{n(n + 1)}{2}$.

If $a = 1$, and if the common difference is 2, then $b = 2n - 1$ and we have n^2. We thus find the results already met with above.

Notice the analogy which exists between the formula $\dfrac{(a + b) n}{2}$ above, and that of the area of a trapezium.

If r is the common difference of an arithmetic progression, this progression can always be represented by

$$a \quad a + r \quad a + 2r \ldots a + (n - 1)r.$$

35. Geometric Progressions.

If a series of numbers

$$2 \quad 6 \quad 18 \quad 54 \quad 162$$

for instance, is such that, on dividing each of them by the preceding one, the same quotient results, these numbers form a *geometric progression* or *progression by quotient*. The constant quotient is the *common ratio* of the progression. It may be said that, in a geometric progression, the relation of one term to the preceding one is constant, and is called the common ratio. In the example given above, the ratio is 3, the first term is 2, and the number of terms is 5.

The numbers 1, 10, 100, 1,000, . . . in the decimal system, form a geometric progression with common ratio 10. The same numbers written in a numeration system with the base B form a geometric progression whose common ratio is B.

The common ratio can be a fraction, as well as a whole number. If it is greater than 1, the terms go on increasing without end; the progression is then called *increasing*. If the ratio is less than 1, the progression is a *decreasing* one, and its terms go on diminishing more and more.

It is interesting to be able to find the sum of the terms of a geometric progression. Let us go back to the example given above

$$2 \quad 6 \quad 18 \quad 54 \quad 162.$$

If we multiply any term by the common ratio 3, we have the following term. If it is multiplied by $3 - 1$ or 2, then we shall have the difference of two consecutive terms:

$$2 (3 - 1) = 6 - 2, \quad 6 (3 - 1) = 18 - 6,$$
$$18 (3 - 1) = 54 - 18, \quad 54 (3 - 1) = 162 - 54,$$
$$162 (3 - 1) = 486 - 162.$$

Adding all these, if the sum is s, we shall have then $s (3 - 1) = 486 - 2$, since all the other numbers cancel; so that $s = \dfrac{486 - 2}{3 - 1} = 242$.

In general, when

$$a\,b\,c \quad \ldots \quad \ldots \quad \ldots \quad k$$

is the progression with common ratio q, of which we wish to find the sum s; if we take it a term further $l = kq$, we shall have

$$a (q - 1) = b - a, b (q - 1) = c - b, \ldots$$
$$k (q - 1) = l - k,$$

and $s (q - 1) = l - a$; $s = \dfrac{l - a}{q - 1}$.

If the progression is a decreasing one, we have $\dfrac{a - l}{1 - q}$, which comes to the same thing.

Geometric progressions play an important part in calculation; they have numerous applications.

Even if the common ratio is not much greater than 1, if the number of the terms becomes rather high, the progressions lead to numbers of an enormous size; these results amaze us to begin with unless we are forewarned. We will quote several instances in the following sections.

It is useful to notice if a is the first term of a geometric progression, q the common ratio, and n the number of the terms, the progression may be written thus

$$a \quad aq \quad aq^2 \quad \ldots \quad aq^{n-1}.$$

36. The Grains of Corn on the Chessboard.

The inventor of the game of chess is not exactly known ; but there exists on this subject an old Hindoo legend which deserves to be remembered.

Enchanted by the new diversion, the monarch, according to this legend, caused the inventor to be brought before him, and invited him to fix for himself the reward which he desired.

" Let your Highness simply deign," responded the man " to order your servants to give me a grain of corn, to be placed in the first division of my chess-board ; 2 on the 2nd, 4 on the 3rd, and continue thus, always doubling, to the 64th division."

The modesty of such a request struck the monarch with astonishment, so we are told, and he gave orders that the request should be satisfied without delay. But he was still more amazed when later he was made aware of the absolute impossibility of fulfilling his commands. In order to have produced the necessary quantity of corn the product of eight harvests would have had to be gathered, always supposing that the whole of the soil of his kingdom had been sown with seed.

The number of the grains of corn required is the sum of the terms of the progression

$$1 \quad 2 \quad 2^3 \ldots 2^{63},$$

which gives $2^{64} - 1$. Here is the number written in the decimal system :

$$18446744073709551615.$$

There are twenty figures in this, as we see. We will

not attempt to read it. The words which we would utter would not convey much meaning to our mind. However we shall presently find others much larger.

37. A very Cheap House.

One of our friends, probably knowing the chess-board story, had a little two-storied house built for him. A flight of 7 steps led from basement to first floor, and the staircase which led to the second floor had 19 steps.

At the end of several years he decided it was time to put his little place on the market, as it was in good order, and had a very pleasant aspect. To the first would-be purchaser Smith made the following proposition :

" I am not at all unreasonable, and, moreover, I really wish to sell ; suppose I offer you the house if you will put a cent on the first of the small flight of steps, 2 on the 2nd, 4 on the 3rd, doubling thus on every step till the end of the staircase is reached. It is really nothing, there are only 26 steps in all."

" Done ! there's my hand upon it," cried Jones, the would-be possessor, beside himself with joy at such a stroke of good luck.

And the next day Jones, having first entertained Smith to a sumptuous meal, set out for the house to count out the required coins on the 26 steps, *i.e.* $2^{26}-1$ cents.

Up to the top of the first flight of steps all went well, and the first few steps of the staircase presented no insuperable difficulty ; but soon his purse emptied much more quickly than he had imagined it possible.

The vendor very obligingly offered to tell Jones the total amount of his debt, so that he would have no need to go up further. " My dear Jones," he cried, " you owe me $671,088.63, but from an old friend like you, I will not expect the .63, I shall be pleased to meet you that far."

Poor Jones' face lengthened remarkably at this

announcement, and ever since he has insisted that each of his children should be taught what a progression means.

He himself is perfectly acquainted with its nature, but his knowledge was rather expensive.

38. The Investment of a Centime.

One of the most important practical applications of progressions is that which concerns *compound interest*. If we put out 100 dollars at interest for a year, at 5 per cent., it brings in 5 dollars. If, instead of touching the 5 dollars, we join them to the 100, that makes 105 which we can put out during a second year, and so on. When the number of years becomes considerable, the growth of capital by this operation of compound interest is absolutely startling.

Suppose, for example, that at the beginning of the Christian era a cent had been put at compound interest at the rate of 5 per cent.; it is calculated that toward the end of the 19th century its acquired value would be more than 200 millions of spheres of pure gold as big as our earth.

We may say, in passing, that such a result shows us the impossibility of an absolute application of compound interest in practice. The enormousness of the amount forbids any exact idea of such a sum.

It will be far better to put a question of this kind : for what length of time must a dollar be put out at compound interest at the rate of 5 per cent. so that the acquired value may be a hundred million dollars?

The answer is 378 years, so that if one of our ancestors, about 1527, in the time of Henry VIII., had conceived the brilliant idea of placing 1 dollar to your credit, at 5 per cent. compound interest, this would have grown to-day to the enormous value of 100 million dollars.

If the same thing had been done in the year 59 of our era, at the rate of only 1 per cent., the same result would have been obtained in 1907, that is to say, the dollar thus placed would to-day be worth 100 million dollars (always supposing no accidents occurred in the interval !).

39. The Ceremonious Dinner.

One evening twelve people had arranged to dine together. Each of them attached great importance to points of etiquette ; now the seating of the party had not been arranged in advance, and a courteous discussion arose at the moment of going to table which, however, did not lead to any result. Some one, as a means of solving the difficulty, proposed that all the possible ways of attacking the problem should be tried ; there would be nothing to do then but choose the one which seemed the best. Accordingly this was done for some few minutes, but they became so mixed up that it did not seem to hold out any satisfactory prospect. Happily, among the guests, there was one, a professor at the college in the town, who was a mathematician. " My good friends," said he, " the soup is going cold. Let us seat ourselves at random; that will be quickest." This wise counsel was followed and the repast was brought to a close amid the greatest cordiality. At dessert, taking up the subject once more, " Do you know," said the professor, " how long it would have taken us to try all the possible ways of seating ourselves round this table, taking no more than just one second to move from one seat to another ? " and, as each kept silence, he went on to say, " Continuing this little game, day and night, without stopping a single moment, we should have been 15 years and 2 months, taking no notice of leap years. You see that the meat would have dried up and we ourselves should all have died of hunger, weariness, and loss of sleep. By all means let us be ceremonious if we wish, but not to excess."

This was absolutely correct; the precise number of

different ways in which 12 people can take their places at a table laid with 12 covers is just 479,001,600 ; more than 479 millions, as you see.

This result is amazing when we reflect that for 2 diners 2 seconds of time only would have been needed ; and even for 4 the trials might have been made in less than half-a-minute.

The enormous numbers we have just mentioned are due to permutations, and the deduction is easy to make.

When several different objects are in question, which can be arranged in various ways indicated beforehand, any particular arrangement adopted is a *permutation* of these objects.

If we deal with two objects a, b and with two different places, the only two permutations possible are $a\,b$ and $b\,a$.

To form the permutations of three objects, a, b, c, we can take the permutation $a\,b$ and join c to it at three different places ; it can come after b, between a and b, or before a.

The permutation $b\,a$ will also give three by joining c to it ; so that we shall have the table of permutations of a, b, c by writing

$$\begin{array}{ll} a\,b\,c & b\,a\,c \\ a\,c\,b & b\,c\,a \\ c\,a\,b & c\,b\,a \end{array}$$

and this gives $2 \times 3 = 6$ permutations.

If we take any one of these permutations, $a\,b\,c$ for instance, and add on to it a 4th letter, we shall have then 4 permutations

$$a\,b\,c\,d \qquad a\,b\,d\,c \qquad a\,d\,b\,c \qquad d\,a\,b\,c$$

and each permutation of 3 letters thus furnishing 4 of 4 letters, the number of permutations of 4 letters will be 6×4, or $2 . 3 . 4 = 24$.

Continuing thus, the number of permutations of 5 letters would be $2 . 3 . 4 . 5$; and generally, that of the

Y   H

permutations of n letters will be $2 . 3 . 4 \ldots n$. This
is often represented by $n!$ or $\lfloor n$.[1]

We can see below how rapidly these $n!$ numbers of
permutations grow when n increases.

n	$n!$
2	2
3	6
4	24
5	120
6	720
7	5040
8	40320
9	362880
10	3628800
11	39916800
12	479001600

Permutations play a most important part in mathematics. They can be used, besides, in various games
and amusements, such as anagrams. Very many papers,
exceedingly learned ones, have been published on permutations. We shall not deal with them here, but may
mention the happy idea of Ed. Lucas, of representing by
a drawing the permutations of several objects. He has
called it *pictured permutations*. To give the pupil a clear
understanding of this idea, suppose that we make on
squared paper a square of n columns, of n rows each;
and, confining ourselves to permutations of 4 objects,
n shall equal 4. There will be a square of 16 divisions.
If we replace the 4 objects a, b, c, d by the 4 numbers
1, 2, 3, 4, the permutation, $c\ b\ d\ a$, for example, can be
written 3 2 4 1, and so with the others. Taking, then, the
first column of the square, we mark the 3rd division,
and shade it, the same for the 2nd division of the 2nd
column, the 4th of the 3rd column, and the 1st of the
4th. The four divisions so shaded thus stand for the
permutation $c\ b\ d\ a$.

Fig. 73 shows us the 24 permutations of 4 objects.

$n!$ is called " factorial n " because it is made up of factors.

To make this easy to understand we give below the table
of permutations which corresponds to the figure, in the
same order.

FIG. 73.

abcd	abdc	adbc	dabc
acbd	acdb	adcb	dacb
cabd	cadb	cdab	dcab
bacd	badc	bdac	dbac
bcad	bcda	bdca	dbca
cbad	cbda	cdba	dcba

н 2

If we consider any one of the squares of Fig. 73 as a chess-board, the shaded divisions represent the positions of castles which cannot be taken by one another, and that applies to all analogous squares. It follows, then, that on an ordinary chess-board of 64 divisions we can place, in 40,320 (8 !) different ways, eight castles so that they cannot take one another. On a board of 100 divisions, ten castles could be placed, under the same conditions, in 3,628,800 (10 !) different ways. We can consult the table given on page 98 for these numbers. Questions of this kind are not easy to answer without the help of permutations. With their aid the solutions are perfectly simple.

You can also ask yourself in how many different ways we can place the cards in a game of piquet ; the answer is 32 !, or those in a game of whist, which will be 52 !, but I do not advise you to try and write these numbers in the decimal system. Try rather to find the time it will take to carry out all these changes, taking a second to do each one. This is a pleasure I am leaving to my readers, or rather to their pupils. But let them not try to write these numbers, even counting them in centuries, for such an effort would hardly tend to the education of one's mind.

40. A Huge Number.

We are raising ourselves, by progressions on the one hand, and permutations on the other, to great heights on the ladder of numbers.

In the hope of returning to more reasonable limits, and thus escaping a feeling of giddiness, ask some one to write down the largest number possible by using three 9's. Generally the answer will be

999,

quite a modest number, indeed, which does not make one's brain whirl in the slightest.

But if by chance you have happened upon a conscientious mathematician, anxious to give you an absolutely correct answer, you will read, by a slight change in the position of the 9's,

$$9^{9^9}$$

This means that we must raise 9 to a power marked by the number 9^9. This last is easily found in a few minutes. Your pupil will certainly give it to you without any hesitation if you do not care about working it yourself. It is

387,420,489,

and this result is very interesting, for you know, thanks to the pupil's answer, that all you have to do to obtain the required number in the decimal system is to make 387,420,488 multiplications.

These are very simple, having only 9 as multiplicator, but their number rather inspires hesitation.

Decidedly, I cannot encourage you to undertake the task. I will only tell you—so that it can be repeated to the pupil, who can verify it later—that the number 9^{9^9}, if written in decimal numeration would have

369,693,100 figures.

To write it on a single strip of paper, supposing that each figure occupied a space of $\frac{1}{5}$-inch, the length of the strip would need to be

1,166 miles, 1,690 yards, 1 foot, 8 inches,

which is farther than from New York to Chicago.

Under the same conditions, to write $10^{10^{10}}$, we would need a strip of paper long enough to encircle the earth.[1]

[1] This remark was made by M. Ch. Ed. Guillaume, in a very interesting article in the *Revue Générale des Sciences* (30th Oct., 1906).

The time that we should spend in writing down the number 9^{9^9}, taking a second for each figure, and working ten hours per day, would be approximately 28 years and 48 days, working continuously without stopping for Sundays and holidays.

To add to your information, I can assure you that the first figure of the number we are seeking is a 4, and that the last is a 9. That leaves us just 369,693,098 figures to find. Perhaps you may think this but a paltry assistance, and I am of the same opinion. However, I hope you will agree with me that the title I have chosen for this section, " A Huge Number," is thoroughly justified.

It is worth noticing that 1^{1^1} is simply 1, that $2^{2^2} = 16$, and that 3^{3^3} is a number of 13 figures,

$$7,625,597,484,987.^1$$

41. The Compass and Protractor.

In the various drawing exercises, which the pupil ought never to have been allowed to discontinue, circles or fragments of such may have occurred which were drawn almost freehand.

---

[1] In spite of the explanations given, several readers have confused the signification of 3^{3^3}, and several have written to me saying that they made the result 19683, and not a number of 13 figures. This arises from a false interpretation of the symbol a^{b^c}, which can be read $\left(a^b\right)^c$ or $a^{\left(b^c\right)}$. This last interpretation is the only reasonable one; for $\left(a^b\right)^c = a^{bc}$. It would be illogical to employ the expression a^{b^c} to represent the more simple one a^{bc}. So 9^{9^9} can only signify $9^{\left(9^9\right)}$ and 3^{3^3} means 3^{27} and not 27^3.

For drawings in which we need a certain amount of precision, the time has come to accustom the pupil to the use of the compass. He must practise by tracing arcs of circles first, then whole circles, in pencil to begin with, and in ink afterwards. He will be shown, following the mode pointed out in all the classical treatises, how to draw perpendiculars to straight lines, to construct angles, and various other figures, etc.

These constructions, when they are required to be exact, ought besides to admit of the use of the *protractor*, which is as simple in its use as the compass, and is of rather similar service in the diverse forms it assumes ; semicircular or rectangular, made of metal, horn, etc.

As to the method of division of the protractor, preference must be given to that in *grades*, where the right angle is divided into 100 grades, and the grade then into tenths and hundredths, etc.

This method of measuring angles was instituted at the same time as the metric system. Then it was abandoned for the old system of degrees, minutes, etc.

In France they are now, and with good reason, returning to the grade method, even in various official lists ; and several important public offices constantly make use of this division into grades. It is most advisable, therefore, to make it, from the outset, quite familiar to the pupils, and to show them the half of a right angle under the name of 50 grades (rather than 45 degrees).

These constructions are to be made, and should be very simple. Indeed, they may often be left to the initiative of the pupil, remembering, however, that it is important to make him execute the same construction by means of different scales. He will conceive thus the notion of figures having the same shape but different size, that is to say, similar figures, without being taught any definition.

The child will perceive very quickly that the scale

chosen to make a construction will cause no change in the angles ; but on the contrary, if a double or treble scale be adopted, all the corresponding lengths will become doubled or trebled. In short, without making any geometrical study as far as the present is concerned, he will acquire a knowledge, born of experience, of many truths, whose proof will be so much the more easily assimilated later.

There are certain other properties useful to know, and certain names useful to retain in the memory, for which the pupil must provisionally take your word, and give you credit. These form the subject of the following sections.

42. The Circle.

The *circle* is the round figure (Fig. 74) that is traced with a compass, one of the points remaining fixed. The

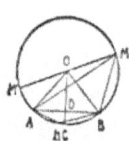

FIG. 74.

point O which is fixed is the *centre ;* the distance from the centre to any point M of the circle is called the *radius.* Twice the radius is the *diameter ;* any straight line, MM', passing through the centre, is a diameter; the length of the segment MM' is double that of the radius. When we take the two points A, B on a circle, that portion of the circle limited to A and B, whether on one side or the other, is called an *arc of the circle.* The straight line AB is a *chord* which *subtends* the two arcs AB. The space between the chord and the arc is called a *segment of a circle.* When the point O is joined to two points A, B of the circle, the angle AOB is called *an angle at the centre.* The angle AMB, whose sides pass through AB, and whose corner is at M on the circle, is an *angle inscribed* in the segment AMB ; this angle is the half of the angle AOB. When we join the centre to the middle C of the arc AB, the straight line OC is perpendicular to the chord AB, which is cut in half at D.

If we take a point N on the circle, below the chord AB on the figure, the sum of the two angles AMB, ANB is equal to two right angles.

When we consider (Fig. 75) a circle and a straight line, this last may be (D_1) outside the circle, or (D_2) may cut it at two points ; this is said to be a *secant* ; or, finally, (D_3) may touch the circle at one point only, in which case it is a *tangent* to the circle. The distance from the centre to the straight line is

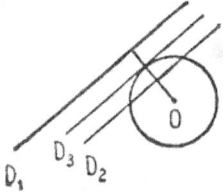

FIG. 75.

greater than the radius for an exterior straight line.

less   ,,   ,,   ,,   ,, a secant.

equal to   ,,   ,,   ,, a tangent.

The point common to the tangent and to the circle is the *point of contact*. The tangent is perpendicular to the radius which is drawn to the point of contact.

In any circle whatever there is the idea of length ; this would be that of an extremely fine thread which would surround the entire ring. Although this general idea lacks precision, it presents a picture, and conveys an impression to the mind ; it will take definite shape later. The length of the circle of which we have just spoken is called its *circumference*.

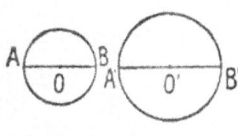

FIG. 76.

If (Fig. 76) we consider any two circles, O, O', the ratio of the circumferences is equal to the ratio of the diameters. This means that the ratio of the length of the circumference to that of the diameter is the same in each of the two circles, and therefore in all circles. This ratio of the circumference to the diameter, which cannot be exactly expressed by any fractions whatever, is greater than 3·14, but less than 3·1416; it is indicated by the Greek letter π. For many ordinary purposes 3·14 is sufficiently near, and 3·1416 will be found exact enough in almost every case when greater precision is needed.

If C is the length of the circumference, and if $D = 2R$ is the diameter, R being the radius, the ratio $\frac{C}{D}$ is then π. This means that $C = \pi D = 2\pi R$. In ordinary practice, $C = 3 \cdot 14 \times D = 6 \cdot 28 \times R$.

This tells us easily the circumference of any circular object, when we know the radius or the diameter, and also how to find the diameter, for instance, of a round tower, of the trunk of a tree, or of a column, when the circumference can be measured with a narrow tape or in any other way.

Direct the children's attention as far as possible to the advisability of doing these exercises on real objects ; do not neglect the opportunity of making them check the approximate value of π which they have used, when, at one and the same time, the circumference and the diameter can be measured.

43. The Area of the Circle.

Just as we acquire, by intuition, a knowledge of the circumference of the circle, so also we feel that the portion of space inside the line has a certain breadth, an area which we should be able to measure. It is found that this area can be obtained by multiplying the length of the circumference by half the radius. And, as we have seen that $C = 2\pi R$, it follows that the area $S = \pi R^2$, or again that $S = \frac{\pi}{4} D^2$. It happens thus that the areas of two circles have between them a relation which is the same as that of the squares of the radii or of the diameters.

Here again, practical examples, as varied as possible, will serve as subjects for exercises on these questions of areas : circular masses of stone in a garden, fountains in parks, the floor of a riding school which is to be covered

with sand, the picture of a round table ; measuring rings
or halos by the difference between two circles, etc., etc.

44. Crescents and Roses.

In the greater number of the treatises on drawing,
models of figures made up of circular elementary forms
are frequently formed which can be traced with the help
of the compass and make interesting exercises.

Merely as specimens, we shall
show here a small number only of
these figures, of which some are
very well known.

If we draw (Fig. 77) a semi-
circle having for diameter BC, and
if we take a point A anywhere on
this line, the triangle ABC is always
right-angled, the angle A being a

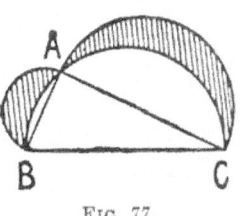

FIG. 77.

right angle.   Now, let us describe two other semicircles
on AB, and on AC as diameters. We shall thus have
two kinds of crescents (shaded
on the figure).   What makes
this figure interesting is that
the sum of the areas of the two
crescents is exactly equal to
the area of the triangle ABC.
This property was known at
the time of the Grecian era, and
the construction that we have
just indicated has become
classical under the name of
the crescents of Hippocrates.[1]

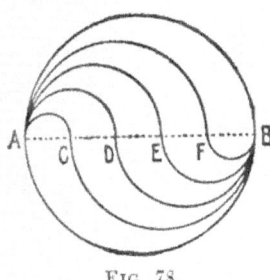

FIG. 78.

Another interesting construction is that shown in Fig. 78.
Let us divide the diameter AB of a circle into five equal
parts by the points C, D, E, F.   On AC, AD, AE, AF
as diameters we draw semicircles above the line ; then

[1] Hippocrates of Chio, Greek geometrician, 5th century B.C.

on CB, DB, EB, FB we draw semicircles below. By means of these circular lines, the circle is divided into five parts which have the same area. Instead of five, any other number n could be taken. It would be sufficient to divide the diameter AB into n equal parts.

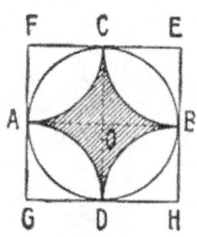

FIG. 79.

In a circle (Fig. 79) let us take two diameters perpendicular to one another AB, CD. Having formed the square OBEC, let us trace from B to C a quarter of the circle of which E is the centre, tracing at the same time three other quarter circles CA, AD, BD; and the whole (the shaded part) forms a sort of star with four points. The area of this star is $(4 - \pi) R^2$, or nearly $0.86 \times R^2$.

If (Fig. 80) we again take two diameters perpendicular to one another, AB, CD, and if we draw the semicircles having for diameters BC, CA, AD, DB, we obtain a rose with four leaves. The area of this rose is $(\pi - 2) R^2$, or nearly $1.14 \times R^2$, the radius being always represented by R.

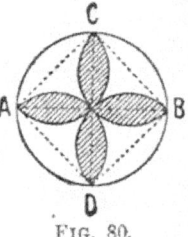

FIG. 80.

Opening the compass to a length equal to the radius, and marking out this length (Fig. 81) successively on the circle at B, C, ... we shall find that it falls once more on the point A after the 6th operation. If, with B, D, F as centres, with the radius R, we describe the arcs of the circles AC, CE, EA, which all pass through the centre, a rose with three leaves will be produced.

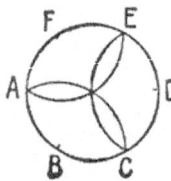

FIG. 81.

By tracing (Fig. 82) with the same construction the six arcs of the circle with the centres A, B, C, D, E, F we obtain a rose with six leaves.

We will limit ourselves simply to these few illustrations, given only by way of example. In practice, they should be constantly varied, and we should impel the child to use his own imagination to form fresh figures. This will follow as a matter of course, for as soon as ever he becomes at all familiar with the use of the compass and other elementary drawing instruments, he will take a pride in forming various figures, and will bestow his time and attention on it.

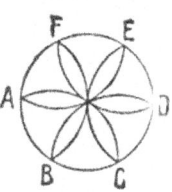

FIG. 82.

45. Some Volumes.

If it is important to determine the areas of surfaces, it is just as necessary, in practice, to ascertain the volumes of bodies. To do this, we must have a *unit of volume*, just as for determining lengths we require a unit of length, and for areas a unit of area. This unit of volume is always the volume of a cube having for its side the unit of length.

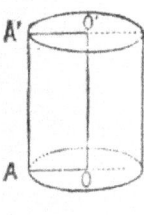

FIG. 83.

Starting from that point, the volumes of a certain number of bodies of regular shapes are found by very simple means.

We intend summarising these now, in such a manner that there will then be no difficulty in solving certain ordinary questions. First of all, let us recall to our minds the bodies that we have already defined, and also point out three others which are frequently to be encountered in ordinary practice.

We have seen what a *cube* is, also a *parallelopiped*, a *prism*, and a *pyramid*.

In all these bodies, we only see straight lines and planes; generally we call them *polyhedra*. In the three others of which we are now going to speak this is no longer the case.

Let us imagine (Fig. 83) that a rectangle AOO'A' turns round its side OO'; it thus makes a body which is called a rectangular *cylinder*. A hat or muff box, a lamp-glass, the inside of a pint pot (sometimes), will show the general form of a cylinder. The two sides OA, O'A' describe two circles of equal radius OA = O'A', which are called the *bases* of the cylinder; OO', which has not moved, represents the distance apart of the planes of the two bases : this is the *height* of the cylinder.

Let there be (Fig. 84) a rectangular triangle AOS, in which O is a right angle, and let this triangle revolve on SO ; this will form a body which is called an upright cone. A sugar loaf, a funnel, a carrot will give a good idea of a cone. The side OA describes a circle which is called the *base* of the cone. The point S, which has not moved, is called the apex. The length SO, the distance of S from the plane of the base, is the *height* of the cone.

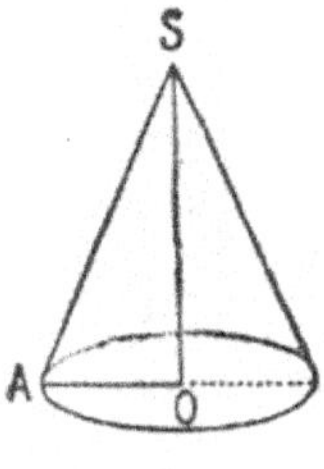

Finally, if a circle turns round on its diameter, the body which this makes is called a *sphere*.

FIG. 84.

The form of the sphere is that of a ball. The centre of the circle is the *centre* of the sphere, and the radius of the circle is the *radius* of the sphere. Any plane which passes through the centre cuts the sphere in a circle. which has the same radius as itself; this circle is called a *great circle*. Any straight line which passes through the centre cuts the sphere in two points equally distant from the centre, and the segment limited by these two points is a *diameter*, of which the length is double that of the radius.

It is well to notice that a cylinder is defined when the radius of its base and its height are given, the same for a cone, and that a sphere is defined when we know its radius.

That being established, we shall have the volume :—

of a cube, by multiplying twice by itself the length of its side. If a measures this length, that gives us $a \times a \times a$ or a^3 : formula, $V = a^3$;

of a parallelopiped, by multiplying the area of the base by the height ; the area B of the base being itself a product ab, if a is a side of the parallelogram of the base whose height is b, the product abh is formed on multiplying by the height of the parallelopiped : formula, $V = Bh = abh$;

of a prism, of which the parallelopiped is only a particular case, by multiplying the area of the base by the height : formula, $V = Bh$;

of a pyramid, by taking a third of the product of the area of the base by the height ; this determination of the volume of the pyramid was given for the first time by Archimedes[1] : formula, $V = \dfrac{Bh}{3}$;

of a cylinder, by multiplying the area of the base by the height. As the base is a circle, with radius r, if the height is h, it follows that the formula is $V = \pi r^2 h$.

of a cone, by taking the third of the product of the base by the height: formula, $V = \dfrac{\pi r^2 h}{3}$;

of a sphere, by multiplying the cube of the radius by the $\dfrac{4}{3}$ of π : formula, $V = \dfrac{4}{3} \pi r^3$.

It is also established that the area of a sphere is equal to four times that of a great circle or $4 \pi r^2$. We can therefore say that the volume of a sphere is equal to its area multiplied by the third of the radius.

Finally, the volume of a sphere can also be determined by the formula $V = \dfrac{1}{6} \pi d^3$, and its area by πd^2, if we call d the diameter.

All these results will be obtained later. They are only

---

[1] Archimedes, an illustrious geometrician. born at Syracuse (287— 212 B.C.).

actually communicated to the pupil in order to make certain practical exercises possible. But do not ask him to overload his memory with all these formulæ.

Put them afresh before his eyes every time that he needs them.

If by constant use they should become fixed in his mind, so much the better. Otherwise, pay no heed to it.

46. Graphs; Algebra without Calculations.

In many of the reviews or journals of to-day we find graphs, figures of which we can make great use for the first mathematical education of children. We must make them understand the signification of these figures, and induce them afterwards to construct similar ones for themselves.

For the most part, the graphs that we are discussing represent variations of meteorological observations, for instance barometric height, temperature, or those of the market price on the Stock Exchange, over a certain length of time. We must also remember that graphs are useful in railway work, in the representation of the movement of trains, and that it is the only practical way of keeping count of them.

But it is above all necessary to notice that by the same means we can represent the variation of any kind of magnitude which is dependent on another magnitude, whether this is time or anything else.

For example, when a weight is hung on an indiarubber thread, this thread grows longer. It will be possible to make a graph which would give us the length of the thread if we know the weight. When we compress a gas its volume diminishes; a graph will tell us what is the volume of the gas when we know the pressure. When we heat the steam from water its pressure increases; a graph will give us the pressure if we know the temperature.

In these various examples the length of the thread depends upon the weight that is suspended; we say it is a *function* of this weight; the volume of gas depends on the pressure; it is a function of the pressure; the pressure of steam, dependent on the temperature, is a function of the temperature. In the preceding examples the height of the barometer, the distance traversed by a train, etc., depended on the period, that is, on the time that has elapsed since a certain fixed period. These were functions of time.

This idea of function is in itself quite natural, quite simple, and a child will easily grasp it if care is exercised in its mode of presentment, by employing as many examples as possible. When a magnitude Y depends upon a magnitude X, and when they are both measurable, the first is a function of the second.

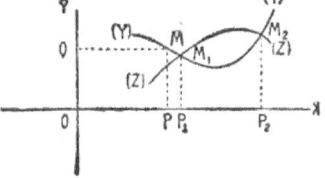

Fig. 85.

The aim of the graph is to bring these functions before the pupil's eyes by means of figures whose construction is always on the same principle. Let us go into this matter.

We will take on squared paper (Fig. 85) two perpendicular lines OX, OY. To show a particular value x of the variable quantity X, we will lay out on OX a length OP, which may be measured by the same number as x, by taking a certain unit of length. To the value x there corresponds a certain value y of Y; we represent this by OQ on OY, taking whatever unit of length we wish; that done, we draw the straight lines PM, parallel to OY, and QM, parallel to OX, which cut each other at a point M; this point represents at once the two corresponding values x, y. By thus constructing points, like M, as many as we like, and joining them by means of

M.   I

a continuous line, the graph showing the variation of the function Y is obtained.

If the quantity X has negative values, and x be one of them, the point P, instead of being to the right of O, will be to the left on OX.

If the quantity Y has negative values, and y be one of these, the point Q, instead of being above O, will be below, on OY.

For any two values whatever which correspond, that is, represent both at once the two points P, Q, there is always one point M and no more.

If the points M which we obtain are not very close together, we can join them by segments of straight lines; we do not even try to picture by a curve the function Y; but the points which show this outline in segments of straight lines will, all the same, give a general idea of the manner in which this function varies.

In algebra—as we shall see later—we do very little but study the functions which can be determined by calculation, and which are called, for this reason, *algebraical functions;* and the fundamental problem of algebra consists in finding values of X, such that two functions Y, Z, of X, become equal to one another. We see (Fig. 85) that, if the two graphs (Y), (Z), of the functions Y, Z were traced, these lines would cut each other at the points M_1, M_2; by taking M_1P_1, M_2P_2, parallel to OY, as far as OX, the lengths OP_1, OP_2 will then give, with the unit adopted to measure the lengths on OX, the two numbers x_1, x_2 which it was required to find.

It is in this sense that we can see that graphs allow us to work by algebra without calculations, and even more than that, since it has been possible to establish graphs for functions which are not algebraical. We ought to add that all the results thus obtained are not rigorously exact. However, in practice, in a great number of cases, if the graphs are carefully made, this approximate result will be all that is necessary. There are many questions to solve which these outlines may be applied with advantage ; besides, they speak to the mind through the intermediary of the eyes, and absolutely place a living representation before the pupil. This is, in itself, a valuable aid to the teacher.

In the following sections some examples will serve still further to enlighten the child's mind as to the construction and the employment of graphs. Their most natural application seems to be in solving the type of problem known under the name of *travelling problems ;* thus we shall especially work with these in various forms.

47. The Two Walkers.

Here we give, under one of its most simple forms, an example to show what the travelling problem is. A pedestrian starts at a given hour, from a given place, at a certain known speed. Some time afterwards, a second, going at a greater speed, starts out in the same direction, following the same route. When will he overtake the first, and at what distance from his point of departure ?

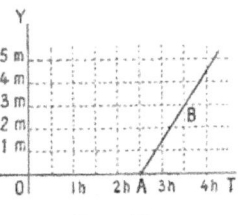

FIG. 86.

To solve this problem, and others of the same kind, we must see how the graph of a pedestrian is made. To do this, on a piece of squared paper (Fig. 86) let us take our two perpendicular straight lines OT (on which we shall mark the time) and OY (on which we shall mark the distances). The point O corresponds to midday, for instance ; let 2 divisions mark an hour, and mark along OT, 1h., 2h., 3h. Again on OY, starting from O, let a division represent a mile, and mark 1m., 2m., 3m. . . . If a man starts at half-past two, with a speed of 3 miles an hour, it will be seen, to begin with, that the graph will contain the point A on OT ; then that at

I 2

half-past three he will have gone 3 miles, which gives the point B ; finally, as the man goes on at 3 miles an hour regularly, the straight line AB will be the required graph. We see that at four o'clock he will be 4½ miles from his starting point, and that the simple outline of the straight line AB shows us at what distance the man finds himself at a pre-determined hour, and at what hour he has gone over a given distance.

We will come back now to our question, and make it more precise by saying that a child starts with a speed of 2 miles an hour, and that a man starts 1 hour after him at a speed of 3 miles an hour. Taking (Fig. 87) the same

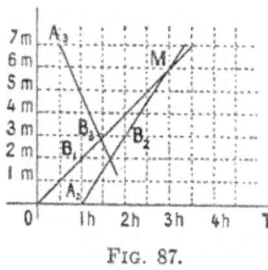

FIG. 87.

units as just now, and counting the time of starting from the departure of the child, then the straight line OB₁ is the graph of the child, and A₂B₂ is the graph of the man.

These lines cut each other at M, corresponding to 3 hours and 6 miles. The meeting will then take place at 6 miles from the point of departure, and 3 hours after the start of the child.

The problem can now be completed by complicating it a little. At a place 7 miles from the starting point a carriage is sent out, going before the two travellers but in the opposite direction. The carriage starts half-an-hour after the child, at the rate of 4 miles an hour. Where will it meet each of the two, and at what time ? A₃B₃ is the graph of the carriage. This straight line cuts OB₁ at the point B₃, showing 1½ hours and 3 miles ; that gives the time and place of meeting with the child. The meeting place with A₂B₂ is at about 1¾ hours (rather less), and at rather more than 2⅓ miles.

Treating this question by ordinary calculation we would arrive at 1 hour 43 minutes as the time, and 2¼ miles as the

distance. It is easy to see, despite the small dimensions
of Fig. 87, that it places results before our eyes
almost exactly correct and perfectly satisfactory in
practice.

48. From Paris to Marseilles.

In the table of trains between Paris and Marseilles,
we have, from the beginning of the year 1905, chosen
the express train No. 1 from Paris to Marseilles and

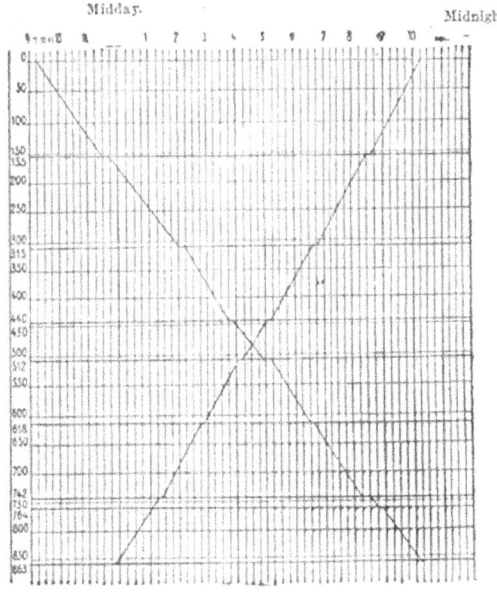

FIG. 88.

No. 16 from Marseilles to Paris (both day trains) to make
the same figure (Fig. 88) show graphs of them. The
time is marked horizontally, the distance (in kilometres)
vertically. The horizontal lines at odd distances repre-

sent various places at various (vertical) distances from either end. On these lines are marked both the time of arrival and of departure. The two graphs cut at a point which tells us when the two trains meet.

We give below the tables showing the hours at which the different stations on the two journeys are reached, in order that a comparison may be made between the variations of the journey and the absolute facts—and this despite the limited dimensions of the figure.

	TRAIN No. 1.			TRAIN No. 16.	
—	Arrival.	Departure.	—	Arrival.	Departure.
Paris	—	9.20 a.m.	Marseilles	—	11.53 a.m.
Laroche .	11.34	11.39	Avignon .	1.20	1.26 p.m.
Dijon	1.59	2.15 p.m.	Valence .	2.51	2.54
Mâcon .	3.31	3.54	Lyons .	4.8	4.14
Lyons .	4.57	5.14	Mâcon .	5.10	5.13
Valence .	6.39	6.42	Dijon .	6.40	6.46
Avignon .	8.16	8.27	Laroche .	8.31	8.36
Tarascon .	8.45	8.53	Paris .	10.20 p.m.	—
Marseilles	10.11 p.m.	—			

We will take advantage of this occasion to notice what a desirable thing it would be if, at any rate as far as the railway is concerned, the habit could be formed of reckoning the hours, starting from midnight, from 0 to 24. This would avoid the use of the words *morning* and *evening,* which prove a fruitful source of mistakes and cause endless confusion. In some civilised countries, this is already done, but not yet in the United States ; let us hope that some centuries hence people will succeed in understanding that it is quite as easy to say " 17 o'clock " as 5 p.m.

I cannot sufficiently impress upon the student the importance of making use of railway guides in the construction of graphs of this nature, and choosing purposely places close at hand, localities which they know already,

at least by name.   Little time will be lost if squared paper
is used and the outlines done in freehand.   They will
then prove very useful exercises.

Single-gauge lines afford matter for most interesting
remarks on the outline of graphs, to show the crossing
of trains which are running in opposite directions.

There is opportunity to notice also how one train going
at a greater speed than another is able to pass it, the
slower one switching into a station, where it remains to
give the quicker train sufficient time to run in front.
There really can be no indication of the thousand interesting details which the construction and observation of
these graphs suggest to us.

49. From Havre to New York.

A long time ago, during a scientific congress, a number
of well-known mathematicians of various nationalities—
some being of world-wide reputation—were dining to-

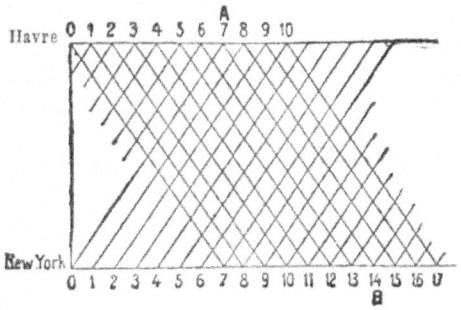

FIG. 89.

gether.   At the end of the meal Edward Lucas suddenly
announced that he was going to lay before them a most
difficult problem.

" Suppose," said he, " (it is unfortunately only a supposition) that each day, at midday, a packet boat starts from Havre to New York, and that at the same time a similar boat, belonging to the same company, leaves New York for Havre. The crossing is made in exactly seven days, in either direction. How many of the boats of the same company going in the opposite direction will the packet boat starting from Havre to-day at midday meet ? "

Some of his distinguished hearers foolishly answered " seven." The greater number kept silence, appearing surprised. Not one gave the exact solution which appears with perfect clearness on Fig. 89.

This anecdote, which is *absolutely true*, instructs us in two ways. To begin with, how patient and lenient we ought to be with those children who cannot immediately take in things which are entirely strange to them. Then, again, the question asked by Lucas shows us the extreme usefulness of graphs as the best method of solving similar problems. Really, if the most ordinary of the mathematicians had had this idea, Fig. 89 would have arranged itself in his brain ; he would have seen it, as it were, with his mind's eye, and would not have hesitated. But they, on the contrary, only thought of the ships about to start, and forgot those on the way—reasoning but not seeing.

It is certain that any boat of which the graph is AB will meet at sea 13 other boats of the fleet, plus the one entering Havre at the moment of departure, plus the one leaving New York at the moment of arrival—15 in all. At the same time, the graph shows the time of meeting to be midday and midnight of each day.

To Lucas also we owe the problem to be found in his *Arithmetique Amusante* under the name of " The Ballad of the Slipping Snail," formulated thus :

" A snail begins to climb up a tree one Sunday morning at six o'clock ; during the day, up to six o'clock in the evening, he gets up 5 yards ; but during the night, he falls back 2 yards.   At what time will he have climbed up 9 yards ? "

This is again a travelling problem (slow travelling this time !).   A bewildered child will answer, " Wednesday morning," which is wrong.   I leave to my readers and their pupils the pleasure of discovering the answer by tracing the graph of the " scramble of the slipping snail."

50. What kind of Weather it is.

Here we give (Fig. 90) two graphs at once, one dealing with barometric pressure, the other relating to temperature, during the last week of the year 1881.   We are borrowing these from a journal (" La Nature "), but taking away, for simplicity's sake, several of the other things shown therein.

Here, we only wish to show how variations of functions, about which we have nothing to go upon but experience, are suitably shown by the graph method.

It also seems interesting to indicate how two different functions can be shown at once on one figure with perfect clearness.

The line which indicates the variations of the barometer is drawn heavily, while the thermometer variations are shown by a dotted line.

To read the barometric pressures, look to the left of the figure, while the figures to show variations of the thermometer (in degrees Centigrade) are to be found on the right.

No confusion is possible at all.   These graphs have rendered the greatest service in Meteorology, and have largely contributed to spread a knowledge of this science, so useful even now, which, although only in its infancy, is progressing by leaps and bounds.

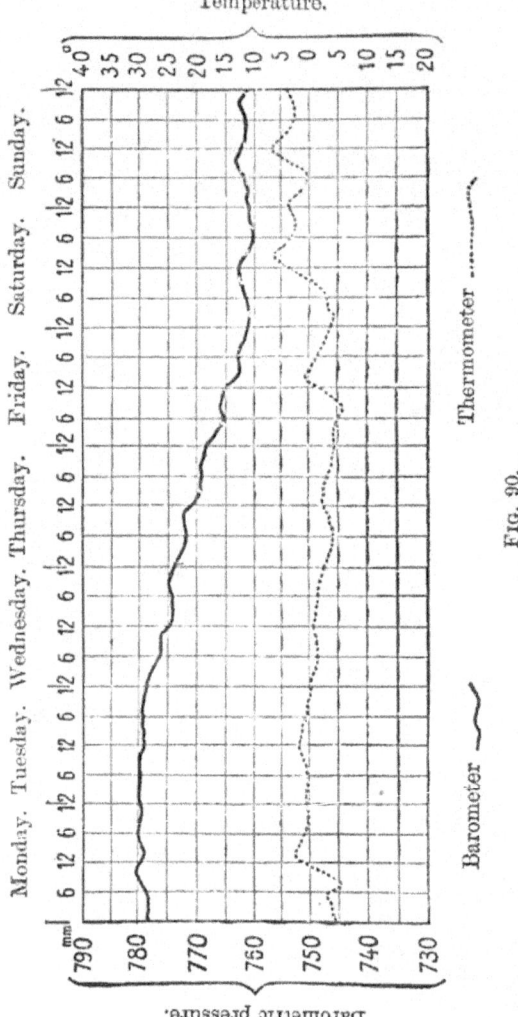

FIG. 90.

51. Two Cyclists for One Machine.

Two cyclists having arranged a certain journey, one of them unfortunately found himself obliged to wait until some necessary repairs could be done to his cycle. However, they decided not to delay their journey, so they arranged it in the following way : that they should start together, one on the cycle, the other on foot ; that at a certain point the cyclist should deposit his machine in a ditch at the side of the road and continue his journey on foot. His companion, on arrival at the spot agreed upon, should then mount the machine, and rejoin the other, when the same thing would be repeated.

FIG. 91.

The programme as arranged was duly carried out, and the last day found them with 20 miles yet to go. When cycling, each traveller goes $7\frac{1}{2}$ miles an hour; when walking, the speed of each is $2\frac{1}{2}$ an hour.

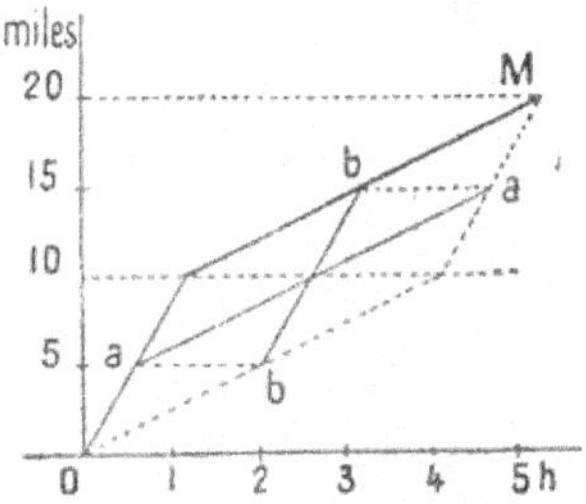

At what point ought the bicycle to be put on one side by the first traveller (no more changing taking place) so that both may arrive at their journey's end at the same time ?

The answer is evident. As each has to go the same distance on foot, as also by cycle, the last change, in order to arrive at their destination at the same time, should take place half-way, that is, 10 miles from the starting point.

Fig. 91 gives us the graphs of the journeys of the two travellers, the first shown by a continuous line, the second by a dotted one. To make the explanation easier, we will suppose that the departure takes place at midday. The cyclist arrives half-way at 20 minutes past 1; there he leaves the machine and goes forward on foot; he finishes his 10 miles, which, at $2\frac{1}{2}$ miles an hour, will bring him to the end of his journey at 20 minutes past 5.

His friend, setting out on foot, does his first 10 miles by 4 o'clock, then he mounts the cycle, and also arrives at 20 minutes past 5.

In short, they have 5 hours and 20 minutes to do 20 miles, which averages $3\frac{3}{4}$ miles an hour. We can see that this mode of locomotion adds sensibly to the speed of a pedestrian; it may be worth something as a hint to two young men who may have just enough money to buy one machine and can share the advantage of it in this manner.

To make this arrangement workable in practice, the changes would have to be made pretty frequently, so that the cycle would not be left long without a rider (unless, of course, the country through which the journey was made was either very deserted, or the people of exceptional honesty). On Fig. 91 we have indicated this variation; supposing that the machine is abandoned at 5 miles, then at 15 from the point of departure OaaM would be the graph of one of the travellers, and ObbM would be that of the other. Here the friends would join half-way, but the cyclist would go on, leaving his companion walking. Graphs take into account all these circumstances.

They would apply equally to two travellers not possessing the same average speed, whether as pedestrians or cyclists. We can thus work problems which lend themselves to calculation without any great difficulty, but which require a knowledge of Mathematics, which we do not suppose the children possess even in the smallest possible degree.

52. The Carriage that was too Small.

Four travellers (Mr. and Mrs. Tompkins and Mr. and Mrs. Wilkins) arrived one morning at the station of X . . ., intending to go for the day to Y . . ., a little village about 31½ miles distant, which they proposed reaching in time for dinner. They had been told that they would be able to hire a motor on arrival which would quickly take them to their hotel, or wherever they were going to dine, along a delightful road. The information proved to be correct, but unfortunately the only available motor would only hold two people and the chauffeur. Its speed was 15 miles an hour.

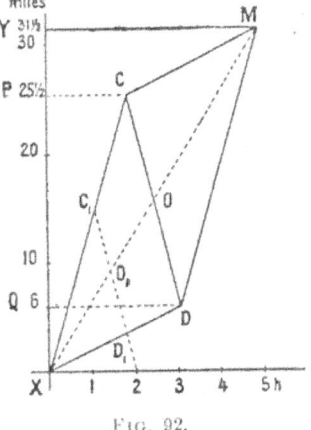

Fig. 92.

My readers can picture the situation. None of the four prided themselves on their powers as pedestrians; they were old and liked to do their modest 2 miles per hour, but no more.

However, it was settled that the Tompkins should start in the motor and that the Wilkins should, at the same time, set out on foot. At a certain distance the motor would put down the T.'s, who would proceed on foot, go back to pick up the W.'s, and carry them to their destination. How would this be managed so that all of them would arrive at the same time, and how long would it take to make the journey?

These questions are not very puzzling, but it is a good thing to be able to solve them.

This problem, except for insignificant changes in the data, has been given at some competitive examinations.

It bears a certain analogy to that of the last section, but is slightly more complicated, owing to the fact that the motor has to come back to pick up Mr. and Mrs. Wilkins.

If we show by P the place on the route where the Tompkins leave the vehicle, by Q the point where it takes up the Wilkins, the four points X, Q, P, Y are arranged in this order : the Tompkins go from X to P by motor, from P to Y on foot ; the Wilkins go from X to Q on foot, from Q to Y by motor. So that they may all arrive together it follows that XP = QY and XQ = PY, which comes to the same thing ; and consequently, as just now, the graph of the journey of the T.'s, and that of the W.'s will form a parallelogram (Fig. 92). But whilst in Fig. 91 the diagonal of this parallelogram was parallel to the axis on which time is measured, the cycle remaining at rest in the ditch, here it will be totally different. The diagonal CD will be no other than the graph of the motor journey when it comes back part way for the W's.

On Fig. 92, XCM shows the journey of the T.'s, XDM that of the W.'s, and XCMD is a parallelogram. These remarks furnish the means whereby the figure can very easily be constructed. To begin with, it is sufficient to draw the two straight lines XC and XD, which is quite simple, since we have the speed of the motor (15 miles an hour) and that of the pedestrians (2 miles an hour). Taking then a point C_1, anywhere on XC (suppose we say the one which agrees with 1 hour and 15 miles), we draw C_1D_1, which represents the graph of the return of the motor if it comes back again to start from C_1 ; this straight line cuts at D_1 the straight line XD. Let us take the middle O_1 of C_1D_1, and, joining X and O_1, the straight line XO_1, produced to the point M, which corresponds to a distance of $31\frac{1}{2}$ miles, will give us the extremity M of the two graphs: draw MC parallel to XD, MD parallel to XC; the parallelogram will be completely drawn, and XCDM will represent the graph of the motor. Then we see on the figure that D corresponds to 3 hours and 6 miles, C to about $1\frac{3}{4}$ hours and $25\frac{1}{2}$ miles, M to $4\frac{3}{4}$ hours, and, naturally, $31\frac{1}{2}$ miles.

Therefore the motor ought to put down the T.'s at a distance of $25\frac{1}{2}$ miles at about 1.45; then come back to pick up the W.'s, finding them at 3 o'clock, 6 miles from the starting point, and bring them on to meet the T.'s at 4.45. An exact calculation would make the arrival 4.42 instead of 4.45, but this, in practice, is of very slight importance.

Summing up the problem, the travellers ought to travel $25\frac{1}{2}$ miles by motor and 6 on foot, and the entire journey is effected in 4 hours 42 minutes.

The average speed is about 6·7 miles an hour, meaning that they arrive at their destination at the same time as if all the journey had been made in a motor with a speed of 6·7 miles an hour. The travellers—both the T.'s and the W.'s—walking 3 hours, would have done 6 miles, and the remainder by motor; as for the motor, it would have run in all $70\frac{1}{2}$ miles, as follows: $25\frac{1}{2}$ onward, $19\frac{1}{2}$ backward, and again forward for $25\frac{1}{2}$ more.

This example, treated thus in detail, will serve as a theme for numerous similar exercises, making use of different data.

53. The Dog and the Two Travellers.

Two travellers are going along a road in the same direction. The first, A, is 6 miles in front of the other, and walks 3 miles an hour; the second, B, walks $4\frac{1}{2}$ miles an hour. One of the travellers has a greyhound, who, at the exact moment of which we speak, runs to the other at a speed of $11\frac{1}{4}$ miles an hour, running immediately back to his master. Having rejoined him, he starts off to do the same thing again, and continues this until the men meet, zigzagging from one to the other. What is wanted is the distance the dog will have travelled up to the moment of meeting.

It appears that the question can be put in two ways, according to which of the men is the dog's owner. In Fig. 93 the time is counted from the moment the dog is let loose. The graphs of the two travellers are OM 6M, and and the point M, which represents the meeting,

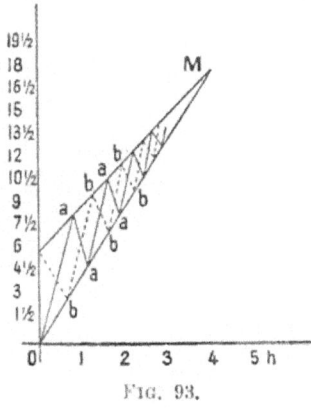

FIG. 93.

corresponds to 18 miles and 4 hours of walking. If the dog belongs to the traveller who is at the back, his graph is O*aa* . . ., a line taking a zigzag course between the journeys of the two men. If, on the contrary, the animal is the property of the man in front, his graph is 6*bb* . . ., a line of the same nature but different from the first. In any case, the dog has never ceased running for 4 hours, and as he goes at a speed of $11\frac{1}{4}$ miles, he will consequently have covered a distance of 45 miles. Whichever hypothesis we take, the result will be the same.

We have taken exceptionally simple instances, to make the explanations very easy. It will be useful to vary them in the exercises which may be given on this subject; for instance, we might suppose that the men start in opposite directions, advancing till they meet.

54. The Falling Stone.

In the travelling graphs that we have seen up to the present time, whether we deal with pedestrians, carriages, railways, or dogs, the distance passed over in a given time, in a second, say, was always the same, and it followed that the graph was a straight line. That is explained by saying that the speed was constant, or that the movement was uniform.

It is not at all the same for a stone that is thrown to a certain height and then is allowed to drop. Experience teaches us, if we do not take into account the resistance of the air, that, at the end of a second, the stone will have fallen nearly 16 feet. At the end of 2 seconds it will have fallen 64 feet, and at the end of the third 144 feet. This shows us that the graph of this movement (Fig. 94) will take the form of a curve, and no longer that of a straight line. This curve will be pretty nearly the one indicated by the figure. It is a fragment of a line about which we shall speak further in a little while, and is called a *parabola*.

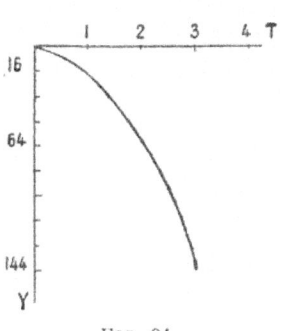

FIG. 94.

Writing the formula $y = 16t^2$, we have the distance y travelled by the stone in its fall when it has fallen during a certain time t, provided that, in measuring the time, we take the second as the unit; then the number y which will be obtained will be a number of feet.

For example, in $\frac{1}{10}$th of a second the stone only falls 2 inches, and, as we have just said, at the end of a second, it has fallen 16 feet. In 10 seconds it will have travelled 1,600 feet. Thus we see that it falls quicker and quicker; in other words, its movement becomes accelerated.

To fall from a height of 900 feet would take a stone about $7\frac{1}{2}$ seconds, always supposing that we do not take into account the resistance of the air. In

M. K

practice this is only very little when we are considering little distances, but it becomes very appreciable
when great heights are in question, and it is a mistake
to think that our graph will then be a correct representation.

55. The Ball Tossed Up.

If we toss a leaden ball into the air it will rise to a
certain height, then fall down. Following the object
with a certain amount of attention, it is not difficult
to prove that the movement becomes slower and slower
during the ascent, while during the descent, on the
contrary, the motion becomes quicker. In the first
period the movement is slackened, in the second it is
accelerated.

Instead of using the hand, let us suppose that we
employ a gun, the barrel of which is placed vertically;
the same effect would be noticed; only we must remember
that the greater the speed at which the ball is launched
the more it will rise, and the more time will elapse before
it falls to earth.

It is interesting to find out various details about
the movement, to know, especially, to what height the
ball will rise; how long it will take to get to this height;
how long it will take in its descent.

When we know the speed a at which the ball has been
thrown, and which is known as the initial velocity, all
the answers to these questions are given by the formula
$y = at - 16t^2$.

To comprehend its meaning, and to make use of it
when necessary, we must know:

1. That the height y is measured in feet;

2. That the initial velocity a is measured in feet
per second; that is to say, that the ball is thrown in
such a manner that if nothing caused a slackening of

speed, it would go on indefinitely travelling *a* feet each second;

3. That the time *t* is measured in seconds.

However simple the calculations may be to which this formula leads, we can follow the movement more easily by means of a graph (Fig. 95).

It has been constructed on the supposition that $a = 64$, that is to say, that the ball is thrown in such a way that it would travel 64 feet per second if nothing happened to oppose its movement. If we construct the straight line OA, which would be the graph of this uniform movement of 64 feet a second, there is a very simple means of obtaining what we wish; going back to Fig. 94, we set out exactly the same heights for 1 second, 2 seconds, 3 seconds, etc., but below OA (Fig. 95) instead of being below OT (Fig. 94).

Another method is to make use of the formula above, $at - 16t^2$, in order to have each value of *y*.

By using any of these, we shall see, on the hypothesis that $a = 64$, that the ball will rise for 2 seconds, that it will

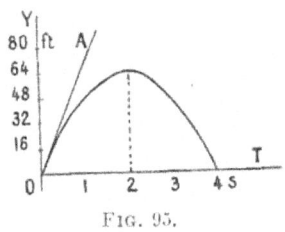

FIG. 95.

reach a height of 64 feet, and that it will take 2 seconds to come down. The line obtained has again the form of a parabola.

In a general way it is found that the time taken up by the descent will be always the same as that of the ascent, and that the height to which the ball will reach is always $\frac{a^2}{64}$, expressed in feet.

Here, as in the preceding section, it is quite understood that no account is taken of the resistance of the air, which, however, for big initial speeds, would have a sensible effect, in both going up and coming down.

56. Underground Trains.

Underground, or tube, railway systems show special working conditions, made necessary by the needs of travelling service in a big city.

To begin with, the stations are very close together; often only some hundred yards lie between them. Besides that, the trains follow each other at short intervals, so that the stop at each station has to take as little time as possible.

Under such conditions a good part of the time needed for the journey from one station to another is employed, on leaving one stopping place, in quickening the speed; then, on approaching the next, in slackening it off; this latter is done by means of brakes, for if the train were brought to a standstill suddenly, an accident might happen.

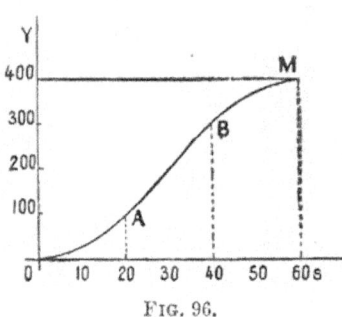

Fig. 96.

Our readers might say that this holds for all railway trains, which is partly true; but as the distances between two stations are sufficiently long, the periods of setting the train in motion and applying the brake count for very little in the whole. This is why, without departing from practical exactitude, we can represent by a straight line the graph showing the journey of a train between two stations.

This journey is, then, interesting, because of these peculiarities, and also of the corresponding graph, which is shown in Fig. 96.

To draw this graph we have supposed two stations distant 400 yards one from the other, a speed through the entire journey of 36,000 yards (about 20 miles) an hour, which means 10 yards a second ; finally it must be admitted that it takes 20 seconds, starting from the halt, to get up full speed ; and equally, of course, 20 seconds to slacken off before the next stopping place.

With these data to hand, corresponding to the working of the journeys we see that a train starting to the next station moves at a rising speed, like a ball which falls faster and faster ; it runs over 100 yards in 20 seconds ; it rolls along then at full speed, at 10 yards a second for 20 seconds, and thus goes 200 yards ; then the brakes are applied, the speed is slackened, the train goes 100 yards in 20 seconds, and stops. It has then arrived at the next station, and it has travelled the distance in 1 minute, or 60 seconds.

The graph (Fig. 96) takes all these circumstances into account; from 0 to A is the period of getting up speed (100 yards in 20 seconds); from A to B, the period of full speed (200 yards in 20 seconds) ; and from B to M the period of slackening speed until finally the train is brought to a halt (100 yards in 20 seconds).

It is sufficient to look at the figure to realise the import-ance of the increasing and the slackening of speed over such small distances. If two stations were distant from each other 200 yards instead of 400 yards, the period of full speed would completely disappear, and it would take 40 seconds for the train to travel 200 yards.

57. Analytical Geometry.

The general idea which underlies the construction of graphs has been shown in section 46, and applied under various forms in the pages following. It consists, as we may remember, after tracing two perpendicular straight lines OX, OY, in setting out on OX a length $x = OP$, on OY a length $y = OQ$, and determining a point M by drawing through P and Q the parallels to OY and OX which cut each other in this point M.

If y is the value of a function of x which we wish to represent, the line obtained by joining all the points M that have been constructed will represent the variations of the function y.

By means of some new illustrations, we are going to find in them everything which is at the base of an important and very useful science, *analytical geometry*, which we owe to the genius of Descartes.[1]

And it is as well to add that without analytical geometry we never could have imagined graphs.

The two straight lines OX, OY (Fig. 97) are called the *co-ordinate axes*; OX is the axis of the x's, or the axis of the *abscissæ*, OY the axis of the y's, or the axis of the *ordinates*.

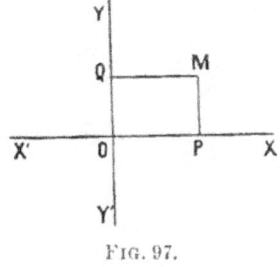

FIG. 97.

$OP = x$ and $OQ = y$ are the *co-ordinates* of the point M; OP is the *abscissa* of M, and OQ its *ordinate*.

A negative abscissa would be set out in the direction OX', a negative ordinate in the direction OY'.

It results from this that if a point, as seen on the figure, is in the angle XOY, its x and its y are positive; if it is in the angle YOX', its x is negative, its y is positive. X'OY', its x and its y are negative; Y'OX, its x is positive, its y negative.

If a point is marked on the plane of the figure, we then know its two co-ordinates. If any two co-ordinates are given, we know the position of the corresponding point.

If the two co-ordinates x, y are not simply any numbers, but are linked by an algebraical relation, that is to say, that one of the co-ordinates being known the other may be deduced from it by a series of perfectly definite calcu-

---

[1] Renè Descartes, a celebrated philosopher and man of letters, born at la Haye, in Touraine (1596—1650).

lations, the positions of M will lie on a line. The algebraical relation in question is the *equation* of the line.

The great general problems with which analytical geometry deals are :—

1. To construct a line, and find out its properties, knowing its equation ;

2. To find the equation of a line, when it has been defined in a precise manner by any means.

Our readers need not be ambitious to learn what analytical geometry really is. But in constructing our various graphs we have done a little of this branch of geometry without even knowing the name of the science ; so it was desirable to profit by the occasion given us to salute in passing the memory of one of the greatest geniuses of whom the world has reason to be proud.

It is since the invention of Analytical Geometry that the study of curved lines has made immense progress, thanks to the fresh resources which this science has brought to bear upon them.

Three of these curved lines, however (and some others also), had been studied in antiquity by Greek geometricians by the help of Geometry alone. The mind is absolutely amazed when we consider what power of intellect, what prodigious efforts of the brain have been necessary for these learned men, of perhaps more than twenty centuries ago, to arrive at the discoveries by which we are now profiting.

The three lines of which we are going to speak are to-day in continual use, even in practice. For this reason we have resolved to say something about them in the sections which follow, not to study them, be it understood, but simply to know what they are, so that the pupil may have an idea of the pleasure and profit he will have when, later on, he will begin to take them seriously.

68. The Parabola.

We have already met this curve, in the graphs of the falling stone, of the ball tossed up in the air, and in a portion of the graph of the underground trains.

The precise definition of the parabola is (Fig. 98) in that each of its points M is equidistant from a given point

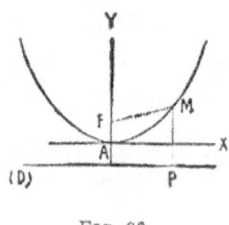

FIG. 98.

F and from a given straight line (D), so that MF = MP. The curve then takes the form shown by the figure ; if from F, which is called the *focus* of the parabola, a perpendicular line is lowered on the straight line (D) called the *directrix*, this straight line FY is the *axis* of the curve, which has the same form on each side of this axis. The axis cuts the curve at A, half-way between the focus F and the directrix. The point A is the *apex* of the parabola.

If AY be taken for the axis of the ordinates, and a perpendicular AX for the axis of the abscissæ, the equation of the parabola would be $y = kx^2$.

59. The Ellipse.

Many of the arches of a bridge take the form of a half-ellipse. When a carrot is cut obliquely with a knife somewhat regularly, the section is an ellipse. If a flat round object such as a coin is held up against a lamp, and the shadow thrown on a piece of white paper, this shadow may also be an ellipse.

Astronomy teaches us that all the planets, and ours in particular, turn round the sun, and in so doing describe ellipses.

The ellipse is determined by this peculiar quality·

that the sum of the distances of any of its points from two given points F, F' is constant; F, F' are the foci of the ellipse. Let us suppose we wish to trace an ellipse on a sandy soil. This can be done by fixing two pegs at F, F' and attaching thereto a cord (of which the length has been given) by its two ends ; this cord is held out by means of an iron spike M ; if this spike is carried over the ground, always keeping the cord stretched out tightly, it will trace the ellipse ; this method is known by the name of " the gardeners' mark."

We see (Fig. 99) that the ellipse is a closed curve ; the straight line AA' is called the *focal* or *major axis* ; the middle O of FF' is the *centre*; the perpendicular BB' to FF' is the *minor axis* ; the curve has a form exactly similar above and below the major axis, to right and left of the minor axis.

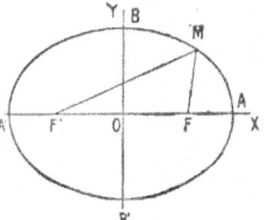

Fig. 99.

The major axis cuts the curve in the two points A, A' ; the minor in B, B' ; the 4 points A, A', B, B' are the apexes of the ellipse. It is easy to see that the constant length MF + MF' is equal to A'A, or twice OA ; this is called the length of the major axis ; the length of the minor axis is BB', or twice OB.

If the two points F, F' were to become one alone, in O, then the ellipse would become a circle, having OA = OB.

Taking OA and OB for axes of the *x*'s and the *y*'s, the equation of the ellipse would be, calling *a* the length OA and *b* the length OB,

$$\frac{x^2}{a^2} + \frac{y^2}{b^2} = 1.$$

The equation of the circle, if *b* becomes equal to *a*, is

$$x^2 + y^2 = a^2.$$

60. The Hyperbola.

Although this curve is also very important, it is not quite so easy to pick out ordinary examples of it as in the case of the two preceding ones. However, if a circular lamp-shade is arranged on a lamp, and then is placed in such a way that the light is below, if we look at the shadow which is cast on a vertical wall by the lower edge of the lamp-shade, we shall see a fragment of a hyperbola.

The hyperbola can be determined by the following peculiar quality : that the difference of the distances from any one of its points to two fixed points F, F', which are called *foci*, is constant.

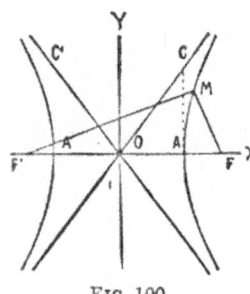

FIG. 100.

As we have seen just now for the ellipse, the straight line FF' (Fig. 100) and the perpendicular OY raised upon the middle of FF' are the axes of the curve. This is of the same form both above and below FF', to right and left of OY. The axis FF' cuts the curve in two points A, A', which are the *apexes*; FF' is called a *transverse axis*; the axis OY does not meet the curve. The segment A'A has a length equal to the constant difference of the distances from a point of the curve to F and to F'.

What we find new here is that the curve, beside being capable of being extended as far as we like, is made up of two parts, of two branches as one may say, completely separated one from the other.

We must note the existence of two straight lines OC, OC', which are called the *asymptotes*, and are such that, by prolonging them, and also prolonging the curve, we shall see the curve and the straight line approach each other, indefinitely, without ever quite running into one. We can easily construct the asymptotes, knowing that the point C is such that CA is perpendicular to FF'. and that OC = OF. If OA = a, OC = c, it follows that $AC^2 = c^2 - a^2$; supposing that AB = b, and taking OA, OY for axes of the x's and the y's, the equation of the hyperbola would be

$$\frac{x^2}{a^2} - \frac{y^2}{b^2} = 1.$$

What we must specially retain in our minds about these very condensed remarks on the three very important curves about which we have just been speaking is that by their aid many and various constructions may be made, and also that they contribute to the acquisition of that manual dexterity which is so necessary in tracing all sorts of geometrical curves. For this purpose the pupil should be encouraged to use, successively or alternatively, squared paper, the usual drawing apparatus, and also outlines in freehand.

61. The Divided Segment.

Let AB be a segment of a straight line ; let it be supposed that it is produced in two directions (Fig. 101) and that M be a moveable point on the straight line AB. If the point M is placed, for example, between A and B, it divides AB into two segments AM, MB, and it is the ratio $y = \frac{AM}{MB}$ of these two segments that we wish to study. It varies evidently according to the position of M.

We will place, to begin with, M at A ; the ratio is nothing, since MA is nothing ; if M is moved from A toward B, the ratio becomes greater ; when M is in the middle of AB the ratio y is equal to 1 ; when M is brought closer to B, y has values which become greater and greater, and it is said that when M arrives at B, the ratio is infinite : or in other words, it is so enormously large that it cannot be expressed in figures.

If, however, M passes a little beyond the point B, AM will be always positive, MB negative, and very small; then y, that is to say, $\dfrac{AM}{MB}$, will be negative and very large; the further M is removed from B (although the ratio will remain negative) the more its size will

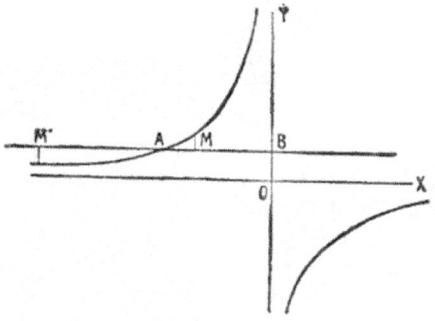

Fig. 101.

diminish, remaining always greater than 1, but approaching more and more nearly to 1.

If now, beginning with the point M at A, we make it move towards the left, the ratio $\dfrac{AM}{MB}$ becomes once more negative; its size is less than 1, and it approaches more and more nearly to 1 in proportion as M becomes distant from A.

Representing, for each position of the point M, the value of the ratio y by an ordinate drawn perpendicular to the straight line AB, we obtain, as a graph showing the variations of this ratio, the curve seen on Fig. 101; this curve is a hyperbola, of which the asymptotes are BY, perpendicular to AB, and OX, parallel to AB, at a distance marked by the unit, and below, that is to say, in the negative direction.

The shape of the figure shows that there are not two points M for which the ratios $\dfrac{AM}{MB}$ can be the same.

As soon as the value y of this ratio is given, with its sign, the precise position of M is determined on the straight line AB.

62. Doh, me, soh ; Geometrical Harmonies.

We have said (Fig. 101) that there cannot exist two such different points M that the ratio $\dfrac{AM}{MB}$ is the same. But a point M being given, we can find another M' from it, and only one, such that the two ratios $\dfrac{AM}{MB}$, $\dfrac{AM'}{M'B}$ may have the same size. Since, then, the signs are contrary, we have $\dfrac{M'A}{M'B} = \dfrac{AM}{MB}$.

When four points M', A, M, B are such that, on a straight line, they may exist thus, we say that they form a *harmonic division*.

The word may appear strange. Before explaining it we are going to write the proportion $\dfrac{M'A}{M'B} = \dfrac{AM}{MB}$ rather differently ; let us call the segments M'A, M'M, M'B, a, m, b. Then $AM = m - a$, $MB = b - m$, and the relation becomes

$$\frac{a}{b} = \frac{m-a}{b-m}, \text{ or, again,} \frac{m-a}{a} = \frac{b-m}{b} ;$$

$$\frac{m}{a} - 1 = 1 - \frac{m}{b} ; \quad m\left(\frac{1}{a} + \frac{1}{b}\right) = 2 ; \quad \frac{1}{a} + \frac{1}{b} = \frac{2}{m}.$$

On the other hand, when we begin the study of sound, we learn that the lengths of a vibrating chord giving the three notes doh, me, soh, which make the perfect major chord, are proportional to

$$1, \frac{4}{5}, \frac{2}{3}.$$

Then the inverse lengths are proportional to

$$1, \frac{5}{4}, \frac{3}{2},$$

or
$$4, \quad 5, \quad 6;$$

and, as $4 + 6 = 2 \times 5$, our three lengths of chords a, m, b will comply with the relation

$$\frac{1}{a} + \frac{1}{b} = \frac{2}{m},$$

written above.

It is this comparison which has led to the name " harmonic division."

More generally, when we have an arithmetic progression of any kind whatever,

$$a \ b \ c \ . \ . \ .$$

and 1 is divided by each of the terms, the result

$$\frac{1}{a} \ \frac{1}{b} \ \frac{1}{c}$$

thus obtained is called a *harmonic progression.*

One of the most remarkable properties of harmonic division, and one which plays an important part in geometry, is the following :—

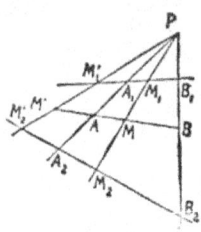

FIG. 102.

Let M′AMB (Fig. 102) be a harmonic division ; if we join the four points which compose it to any point P, and if we cut the four straight lines PM′, PA, PM, PB by any straight line whatever, we shall still have a harmonic division.

Thus on the figure, $M'_1A_1M_1B_1$, $M'_2A_2M_2B_2$ are harmonic divisions. The system of 4 straight lines PM′, PA, PM, PB is called a *harmonic sheaf of lines.*

63. A Paradox: 64 = 65.

In mathematics we often meet with paradoxes, that is to say, we obtain results which we think we have worked out correctly, which are, however, obviously wrong.

Any paradox unexplained is dangerous, because it throws the pupil's mind into a state of doubt and confusion.

When a paradox is explained, on the contrary, it is instructive, because it draws attention to a snare, and shows the illusions of which one may be the victim. Sometimes it is incorrect reasoning, sometimes it is a construction too loosely made, which leads to a flagrant absurdity.

But if paradoxes, properly explained, have thus their place in the *teaching* of Geometry, it is wise to adopt prudent reserve in this matter in *elementary instruction* on the subject. With this last, of course, there is no question of going deeply into things, and they are only indicated, and the pupil just touches them, as it were, with the tip of his finger.

It is this which has decided me to refrain up to now from presenting any question of this kind. Having, however, arrived almost at the end, I see nothing unwise, indeed rather the contrary, in making just one exception which is very well known at the present time. This we might leave the pupil to seek himself. It is hardly likely he will hit upon the best way, and it will be best to come to his assistance without allowing him to become dispirited.

We will take (Fig. 103) a square of 64 divisions on a piece of squared paper and gum this on cardboard. This done, the lines marked on the figure should be traced, and the square will be found to be split up into two rectangles having 8 sides of divisions for base, and heights of 5 and 3 sides; then the large rectangle will be split up into two trapeziums, and the small one into two triangles.

Cut up the cardboard with a penknife or a pair of scissors, by following the three traced lines, which will then give us the four pieces, A, B (trapeziums) and C, D (triangles).

The four pieces must be arranged as is shown in the second part of the figure. We have a rectangle which shows 5 columns of 13 divisions each; we see then 5 × 13, or 65 divisions, with this second arrangement; in the square there will be only 8 × 8 or 64 divisions. These two different results have been obtained with the same pieces of cardboard. This is enough to make us imagine that our heads have become bewildered, seeing that 64 = 65.

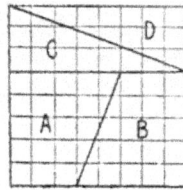

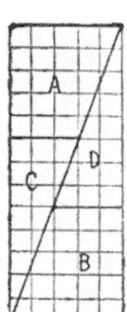

Fig. 103.

The explanation is not very complicated once it is put plainly before the pupil, but it needs some reflection. Looking at the long diagonal of the rectangle, in the second part of the figure, we ask ourselves if it is really a straight line. It is made up of two parts: the hypotenuse of the rectangular triangle C, and the side of the trapezium A. According to the outline, the slope of the hypotenuse on the large side is $\frac{3}{8}$; that of the side of the trapezium is $\frac{2}{5}$. If these two fractions were exactly equal, we should have a straight line. But they are $\frac{15}{40}$ and $\frac{16}{40}$; the first is a little less than the second, and what appears to be a straight line is really a quadrilateral, very thin and very much drawn out, which corresponds to the area of the added division. The union seems exact, but really it is not quite perfect.

If we took a square of $21 \times 21 = 441$ divisions, dividing the side into 13 and 8, we would apparently have, by a similar construction, $441 = 442$.

In that case the two fractions whose equality would be necessary to make a perfect match would be $\frac{8}{21}$ and $\frac{5}{13}$, they would differ only by $\frac{1}{273}$, so that practically the agreement would be perfect.

64. Magic Squares.

If the numbers 1 to 9 are written in the divisions of a square in the following manner,

$$
\begin{array}{ccc}
4 & 9 & 2 \\
3 & 5 & 7 \\
8 & 1 & 6
\end{array}
$$

we can prove that, on adding the numbers contained in a line, in a column, or in either of the two diagonals, the result is always the same:— $4 + 9 + 2 = 3 + 5 + 7 = 8 + 1 + 6 = 4 + 3 + 8 = 9 + 5 + 1 = 2 + 7 + 6 = 4 + 5 + 6 = 2 + 5 + 8 = 15$.

Such a figure is what is called a magic square of 3; the sum 15 is the constant magic sum; if we take 1 away from each figure, which then reads

$$
\begin{array}{ccc}
3 & 8 & 1 \\
2 & 4 & 6 \\
7 & 0 & 5
\end{array}
$$

there would still be a magic square, but the constant would be 12 instead of 15.

Taking the numbers 0, 1, 2, . . . 24, which would fill a square of 25 divisions we would find a magic square of 5; the constant would be 60.

M.　　　　　　　　　　　　　　　　　　　L

The following is an example by means of which it can
be proved that all the requisite conditions have been
properly fulfilled :—

$$
\begin{array}{rrrrr}
0 & 19 & 8 & 22 & 11 \\
23 & 12 & 1 & 15 & 9 \\
16 & 5 & 24 & 13 & 2 \\
14 & 3 & 17 & 6 & 20 \\
7 & 21 & 10 & 4 & 18
\end{array}
$$

and, moreover, if the square is cut by a vertical straight
line between any two of the columns, and if the two pieces
are interchanged, we still have a magic square. Supposing that the square be cut in two by a horizontal
straight line, and the two pieces interchanged, still again
we find a magic square.

Ed. Lucas has given the name " diabolical," to squares
which possess this property.

Magic squares have furnished food for much reflection.
Although they appear to be just a simple game, they give
rise to questions which present great difficulties, and even
the most illustrious mathematicians, Fermat amongst
others, have not disdained to occupy themselves with
them.[1]

We can hardly ignore the existence of these figures
so have pointed them out by way of curiosity.

65. Final Remarks.

If I had to initiate children into the knowledge of
things mathematically essential, such as we have dis-
cussed, this is about what I would say to them at the end
of our course :

" You are going to begin your instruction in mathematical matters. According to your natural dispositions,

[1] One of the most remarkable works published on this question in
our time is that of M. G. Arnoux : *Arithm tique graphique ; Les Espaces
arithmétiques hypermagiques ;* Paris, Gauthier-Villars, 1894.

according to the direction which you will be called upon to follow later in life, this instruction will be more or less of an extended nature ; but, within certain limits, it will be necessary for each one of you.

" Up to the present you have studied nothing, but you have learnt a certain number of useful things, by way of amusement. If you have made any effort, it has been purely a voluntary one on your part, nothing has been required from you, and, particularly, nothing from your memory.

" Before knowing how to read or write, you have been able to make up numbers with the aid of various objects, and to do several simple problems. When it has been possible to employ figures, the practice of calculation has become more easy for you. Thanks to the custom of carrying you back to the objects themselves, and of not only considering the figures which translate them, you have very early arrived at the idea of negative numbers, and become quite familiar with it. Some notions of geometry, found out, but not demonstrated, have been sufficient to begin to make you see the close bond which unites the science of numbers to that of space.

" You have not made a study of fractions more than any other study, but you know what a fraction is, and you have a fair grasp of the calculations which belong to it.

" By progressions, first in simple form, then somewhat more generalised, you have been led to the idea of enormous numbers. Other large numbers appeared before your eyes when you saw what a permutation meant.

" With some practical notions of geometry and drawing at the same time you have succeeded in grasping the construction and the use of graphs, and applying your knowledge especially to questions of movement. You have thus arrived, as it were, at the door of analytical geometry ; you have, at least, perceived the form of the

L 2

three principal curves that analytical geometry permits us to study more deeply, but which the ancients already knew.

"Whether of all these ideas much or little remains in your memory, you are certain to have retained something. You have at the same time acquired, without any doubt, certain habits of mind which are now going to prove of the utmost value to you.

"Henceforward you have not to do with play but with work. You ought to subject yourself to intellectual efforts, perhaps also to some efforts of memory. They will be the less formidable because up to now your forces have been husbanded, and you know many more things than other children of your age who have been subjected to a sort of torture, that of forcing them to retain words in their minds without understanding anything about them.

"In the majority of the objects of your studies in the future you will find things cropping up that you knew of old; any trouble which novelty brings to you will be soon wiped out. Do not think, however, that you will never meet any difficulties; you will find them, but knowing that they are only in the nature of things, that it is necessary to surmount them in order to arrive at interesting and useful results, you will find that you possess the necessary courage. In play you have acquired ideas, and your studies in future will thereby be facilitated. In your work henceforward you are going to make the most of what you know; you will exercise your reason; you will augment the extent of your knowledge. But this work, even if it is no longer play, will not prove to be a burden! You will find a pleasure in it, knowing it to be useful; little by little it will become a necessity of your life; it will not only be easy, but necessary.

"In case of doubt, besides, you will have teachers who will be guides to you; but do not ask anything more from them. Personal, untrammelled effort can alone give good results. You have unconsciously acquired the habit in the games of your childhood. Now it is for you to make the most of it by bringing to the task of acquiring knowledge all the patience, the energy, the determination that you have held in reserve ! ''

Such is about the substance of what ought to be said to the child at the end of these introductory occupations, on the eve of undertaking his studies. To make him grasp these ideas you must not deliver a lecture, but you must explain it, if necessary, in ten or twenty talks. The teacher will have to draw from them the material to light the pupil along the new path that he is called upon to follow.

The introductory process, to my mind, ought to be specially carried out in the home. But even when, from any reason whatever, personal or social, this cannot be, the father and mother ought to remember that their first duty is to associate themselves with the evolution of the child's brain, and to be at any rate a help to the teacher, even if, from any cause, they themselves have been unable to fill the post.

And as, once the introductory stage passed, that of instruction begins, the duty of the parents becomes more important still (if that is possible); their responsibility is heavy, for, whether for good or evil, the whole destiny of their child may be influenced, according to the decision of the father and mother.

It is to these that I turn my attention for the moment, to give a few words of advice—in my opinion, at least, good advice—of which each can take any portion that is likely to be useful.

To begin with, we all agree on one point—that an introduction to the science of mathematics is *indispensable* to any child, without distinction of fortune, of social position, or of sex; but I also maintain that, without any distinction or reserve, mathematical *instruction* is equally indispensable.

Women have need of it just as much as men ; every-day

life, domestic economy, no less than the manufactures and arts whose applications have to do with our existence, require from us all a knowledge of the science of size and space.

Here an objection presents itself which I have refuted a hundred times already, but will discuss once more with my readers. Parents say to me, "Has my child any gift for mathematical study ? If he is not so gifted, is it not losing his time to direct his studies in that particular channel ? We do not intend to make him into a mathematician."

This is all very well. But when you taught the same child reading and writing, did you ask yourself whether he had any gift for these branches of study ? When you inculcated the first principles of drawing, did you think he was intended to become a great painter ? No one doubts the necessity that exists for each man and woman to learn how to express his or her ideas correctly in the mother tongue ; and when that is achieved, surely we do not imagine that each of them is destined to become a Shakespeare or a Milton.

No more in mathematics than in other subjects does instruction *make* learned men ; there is no question of making them ; but there exists in everything a general groundwork of useful knowledge, which is necessary and at the same time easy for everybody to acquire whose brain is not in any way defective.

The whole of this knowledge on various subjects can be acquired, thanks to the preliminary introduction, in much less time than is given up to it in the ordinary course of teaching.

This literary knowledge, as far as our subject is concerned, is pretty nearly represented by elementary mathematics. Any child, whether gifted or not in any special manner, can assimilate the whole of this knowledge, just as he can learn to read and write correctly, if not **elegantly.** If he has an inborn taste for mathematics, he will continue his studies in that direction; if he is literary by temperament, he will write. Teaching has never made learned men or artists, its aim should be the preparation of men's minds.

Then let there be no hesitation on this point. Your child ought to acquire the fundamental notions of mathematics necessary for everybody.

We should always bear in mind the apt and suggestive remark of M. Emile Borel:

"A mathematical education at once theoretical and practical can exercise the happiest influence over the formation of the mind."

I am content to leave you under this impression.

INDEX

THE COUNTRY LIFE PRESS, GARDEN CITY, N.Y.